T0320109

CD files for this title can now be downloaded

from the title's home page on Wiley's website www.wiley.com

by entering the ISBN

9780470387405

PREPARING FOR OSHA's VOLUNTARY PROTECTION PROGRAMS

PREPARING FOR OSHA's VOLUNTARY PROTECTION PROGRAMS

A Guide to Success

BRIAN T. BENNETT
NORMAN R. DEITCH

A JOHN WILEY & SONS, INC., PUBLICATION

Copyright © 2010 by John Wiley & Sons, Inc. All rights reserved

Published by John Wiley & Sons, Inc., Hoboken, New Jersey
Published simultaneously in Canada

No part of this publication may be reproduced, stored in a retrieval system, or transmitted in any form or by any means, electronic, mechanical, photocopying, recording, scanning, or otherwise, except as permitted under Section 107 or 108 of the 1976 United States Copyright Act, without either the prior written permission of the Publisher, or authorization through payment of the appropriate per-copy fee to the Copyright Clearance Center, Inc., 222 Rosewood Drive, Danvers, MA 01923, (978) 750-8400, fax (978) 750-4470, or on the web at www.copyright.com. Requests to the Publisher for permission should be addressed to the Permissions Department, John Wiley & Sons, Inc., 111 River Street, Hoboken, NJ 07030, (201) 748-6011, fax (201) 748-6008, or online at http://www.wiley.com/go/permission.

Limit of Liability/Disclaimer of Warranty: While the publisher and author have used their best efforts in preparing this book, they make no representations or warranties with respect to the accuracy or completeness of the contents of this book and specifically disclaim any implied warranties of merchantability or fitness for a particular purpose. No warranty may be created or extended by sales representatives or written sales materials. The advice and strategies contained herein may not be suitable for your situation. You should consult with a professional where appropriate. Neither the publisher nor author shall be liable for any loss of profit or any other commercial damages, including but not limited to special, incidental, consequential, or other damages.

For general information on our other products and services or for technical support, please contact our Customer Care Department within the United States at (800) 762-2974, outside the United States at (317) 572-3993 or fax (317) 572-4002.

Wiley also publishes its books in a variety of electronic formats. Some content that appears in print may not be available in electronic formats. For more information about Wiley products, visit our web site at www.wiley.com.

Library of Congress Cataloging-in-Publication Data:

Bennett, Brian T.
Preparing for OSHA's voluntary protection programs : a guide to success / Brian T. Bennett.
p. cm.
Includes bibliographical references and index.
ISBN 978-0-470-38740-5 (cloth)
1. Industrial hygiene—United States. 2. Industrial safety—United States. 3. United States. Occupational Health and Safety Administration. I. Title.
[DNLM: 1. Occupational Health.]
RC967.B45 2010
616.9'803—dc22 2009017257

10 9 8 7 6 5 4 3 2 1

CONTENTS

FOREWORD

PREPARING FOR OSHA'S VOLUNTARY PROTECTION PROGRAMS: A GUIDE TO SUCCESS

In my view, OSHA's Voluntary Protection Programs (VPP) are the most successful programs of their kind ever developed by a regulatory agency in the United States. The few OSHA resources dedicated to VPP are leveraged to improve the safety and health of millions of workers, either in those sites maintaining VPP status, or in the many more workplaces that aspire to be recognized as among the country's safest.

The reason for this success is simple. OSHA's Voluntary Protection Programs encourage all employees across the responsibility spectrum (hourly workers through senior management) to focus on workplace health and safety, and to take ownership of sustained OSHA compliance and continuous workplace hazard recognition and control. All participants in VPP realize that achieving "zero workplace injuries and illnesses" is not merely a stated goal or mission, but also actually possible through proper management commitment, worker involvement, management systems, and continuous improvement.

While the supporting empirical data may not yet be fully developed, safety and health professionals, workers, middle managers, and Chief Executive Officer (CEOs) recognize the value OSHA's Voluntary Protection Programs bring to hazard recognition, hazard control, and achieving sustained OSHA compliance. A CEO once told me that whenever a decision must be made on where to grow a business or expand an existing product line, he will, without reservation, choose a VPP site for such an investment over one that has not achieved the VPP level of performance. VPP sites represent the right model for the future, and are wise investments in the 21st Century.

The transformation that must take place at many sites in order to qualify for VPP not only creates a compliant and safer work environment, but also assures that the site will have a culture that is focused on worker participation, ownership, solving problems, and continuous improvement. Leaders in high-performance organizations realize that the key to achieving superior health and safety performance is to engage employees at all organizational levels. Transformational leaders also know that worker participation, ownership, problem solving, and continuous improvement transcend workplace safety and health, and impact morale, quality, productivity, and business focus.

Within the pages of this book, the reader will discover exactly how to achieve the higher performance level required to qualify for VPP recognition. Authors Brian Bennett and Norman Deitch have a unique perspective on what works and how to effectively establish the necessary management commitment, worker involvement, management systems, and continuous improvement in order to achieve success. In my 30 years of professional practice, I have seen many organizations struggle to find the right mix of corporate statements, organizational structure, policies, practices, and training techniques in order to improve workplace health and safety; by following the VPP process and the guidance contained in this text, an organization can find its way to superior health and safety performance.

The journey towards qualifying for VPP may be a long one for many sites, since a site's safety and health culture is always complex, comprised of personal values, beliefs, perceptions, and attitudes influenced and shaped by governing management systems and practices. The process of modifying personal actions and behaviors and establishing sound governing management systems, all of which are required in VPP, takes a considerable amount of time to accomplish. Such a journey is highly worth the effort, as demonstrated by reduced pain and suffering and needless costs, and increased morale, quality, and productivity.

JOHN L. HENSHAW
Assistant Secretary of Labor
2001–2004

PREFACE

We have spent the vast majority of our professional careers in one of the noblest of callings – the protection of our co-workers safety and health. In our efforts to provide the highest level of safety and health, we have embraced the concepts of OSHA's Voluntary Protection Programs. We have tried to capture our 40+ years of collective experience in working with the VPP and provide a comprehensive guide on how to achieve VPP recognition. This book should be useful to both those worksites pursuing VPP recognition, and to existing VPP worksites as a refresher and a source of new ideas.

Chapter 1 is a comprehensive overview of the history of the VPP, along with a description of the various programs.

As with any new initiative, management support must be garnered before the VPP process can move forward. In Chapter 2, we present the business case for pursuing VPP recognition. There are many reasons to pursue VPP recognition, with the primary factor being a significant reduction in workplace injuries and illnesses. Beyond that, there are many other benefits that will help to make your business more efficient, effective, and profitable.

In Chapter 3, we draw from our experience as both preparers and reviewers of VPP applications to provide some insight to preparing an acceptable VPP application. Included is an application which you can customize and use as the basis of your application.

Chapters 4, 5, 6, and 7 cover the four elements, or cornerstones, of the VPP. We provide some tips and techniques you can use to strengthen your safety and health management system. Many of the ideas presented in these chapters are "best practices" observed at VPP worksites.

Chapter 8 will provide some guidance on how to prepare for and conduct the OSHA VPP on site evaluation. Once again, we have drawn from our experience of conducting hundreds of VPP on site evaluations to provide you with some inside information on how to ensure your evaluation runs smoothly.

Chapter 9 includes information on all of the post evaluation activities, including the final report, closure of action items, achieving Merit goals, and celebrating your success as you join the ranks of VPP worksites.

The work is not done once you achieve VPP status; in fact, many say that is when the work really begins! Chapter 10 will cover the requirements and process for your VPP re-approval.

Finally, Chapter 11 provides a wealth of resources that are available to you as you pursue and maintain VPP recognition. The authors encourage all worksites pursuing VPP status to join the Voluntary Protection Programs Participants Association (VPPPA) and take advantage of the services they make available to assist you in achieving your goal of VPP recognition.

Included with this book is a compact disc. This CD includes valuable supplemental information, such as the *Federal Register* notices mentioned in the text, templates of VPP documents, samples of required VPP submittals, and the OSHA VPP Policies and procedures Manual.

Through the years our association with the VPP has had many rewards and led to many close personal relationships. Our peers have become friends, and they have provided the impetus for us to strive for excellence in safety and health. We would be remiss if we did not recognize all the VPP worksites, the members of the National and Regional Chapters of the VPPPA, the professional staff at the VPPPA National office, the staff at OSHA's National and Regional offices, and especially the members of the Region 2 VPPPA Region 2 Board of Directors for their friendship and all of their efforts to improve worker safety and health.

Brian T. Bennett
Norman R. Deitch

Woodbridge, NJ
Langhorne, PA

It is only through the continued love, support, and sacrifice of my wife Sharon and son Tim who continue to provide the inspiration for me to take on additional challenges that this book has come to fruition. Although not listed as authors, they are very much a part of this work.

Brian T. Bennett

Thank you to my wife and partner Shelly and our family for their support and understanding of the work that led to this book.

Norman R. Deitch

1

WHAT ARE THE VOLUNTARY PROTECTION PROGRAMS?

The Occupational Safety and Health Administration's (OSHA) Voluntary Protection Programs (VPP) are a group of programs that provide significant focus on the importance of safety and health in the workplace. It is a recognition program for those workplaces that have gone to extra lengths to not only meet the OSHA guidelines for effective safety and health management systems but also to exceed the OSHA standards wherever feasible. These workplaces have recognized that having a strong management commitment to safety and health includes involving active employee participation in all safety and health activities.

Having described what the VPP is, albeit in a very brief summary, no detailed description of the VPP would be complete without a brief history of workplace safety and health in the United States and a brief history of OSHA. We have come a long way in safety and health since the beginnings of the industrial revolution.

In the dark ages for workplace safety and health, workplace injuries and fatalities were considered part of the cost of doing business. Workplace illnesses were not even a major consideration. From the late nineteenth century to the early twentieth century there were no workplace safety laws or rules. It was the obligation of injured workers or their families to sue their employers for any remedy resulting from a workplace injury or fatality. To be successful in their suits, the plaintiffs had to demonstrate that the employer was at fault. This was hard to prove and most employees were unsuccessful in their suits.

Adding to this problem was the fact that there were no workplace safety standards for equipment or production methods. Equipment manufacturers did not build safety devices into their equipment. Even after OSHA was formed and started issuing

Preparing for OSHA's Voluntary Protection Programs. By Brian T. Bennett and Norman R. Deitch
Copyright © 2010 John Wiley & Sons, Inc.

citations for machine guarding violations, a typical employer defense was that "if the manufacturer thought their equipment was unsafe they would have designed it with a guard." Of course, the typical OSHA response was that the manufacturer was not using the equipment, and it was the employer of the operator that was responsible for the employee's safety, not the manufacturer.

Compounding the issue was the fact that manufacturers were pushing for more productivity. Rather that considering the workers as valuable assets, they were typically considered as the equivalent of a raw material. There was a large immigrant population that was looking for work that represented a boundless resource pool of new workers. The tide began to change in the early 1900s. New York passed the first state workers' compensation law in 1910 that defined the compensations for specific injuries at a predetermined amount. This also made it easier for the employee to receive injury compensation and resulted in additional expenses for the employer. Other states followed New York's example, and by 1921 several other states passed their own workers' compensation laws.

The resulting increase in the cost of workplace injuries caused concern among employers, and they began to pay more attention to reducing injuries and fatalities. They worked with equipment manufacturers to design safety devices for their equipment and began providing their employees with protective devices such as personal protective equipment including hard hats and safety glasses. In 1913, a group of employers formed the National Safety Council to address workplace safety issues collectively. The same year the U.S. Department of Labor was formed. This may be considered as the start of the modern safety revolution in the United States.

These efforts to focus on reducing the number of workplace injuries and their related costs were greatly successful. From 1901 to 1944 the fatality rates decreased from 0.40 to 0.13 per million work hours. For the same period, the injury rates decreased from 44.1 to 11.7 per million work hours. Between 1912 and 2005, unintentional injury deaths per 100,000 population were reduced 51% [after adjusting for the 1948 sixth revision to the World Health Organization's (WHO) International Classification of Diseases and other causes of unintentional death] from 82.4 to 38.1. The reduction in the overall rate during a period when the nation's population tripled has resulted in significantly fewer people being killed due to unintentional injuries than there would have been if the rate had not been reduced.[1]

These successes do not tell the whole story however. In addition to the individual losses, one must also consider the major workplace tragedies that the United States has experienced. These include:

- Over 5000 American deaths during the construction of the Panama Canal
- The 148 workers killed in the Triangle Shirt Waist Factory in New York City on March 26, 1911, mostly young immigrant women and girls
- The 25 workers killed on September 3, 1991 while working at a chicken processing plant in Hamlet, North Carolina
- And, most recently, the 15 workers killed and over 100 injured as a result of the March 25, 2005, explosion at the BP refinery in Texas

These are just a few of the catastrophes that have occurred. Clearly, we still have a long way to go!

FORMATION OF OSHA

The U.S. Congress passed the Walsh–Healy Public Contracts Act in 1936. One section of that act defined specific requirements for workplace safety and health for all federal public contracts. That eventually led to the passage by President Richard Nixon on December 29, 1970, of the Occupational Safety and Health (OSH) Act, Public Law 91-596. The purpose of that act was to "assure safe and healthful working conditions for working men and women." The Department of Labor created the Occupational Safety and Health Administration (OSHA) on April 28, 1971, to administer the OSH Act.

Although OSHA is primarily a regulatory agency with responsibility for enforcing the numerous safety and health standards using inspections, citations, and penalties, the act also empowers the Department of Labor through OSHA to encourage "employers and employees in their efforts to reduce the number of occupational safety and health hazards at their places of employment, and to stimulate employers and employees to institute new and to perfect existing programs for providing safety and healthful working conditions."[2] The act also encourages "joint labor–management efforts to reduce injuries and disease arising out of employment."[3]

OSHA has acted on these empowerments to create a group of voluntary compliance partnership programs to assist those workplaces that have expressed a desire to demonstrate their commitment to providing a safe and healthful workplace and to work with rather than against OSHA toward that goal. This is significant because it demonstrates that OSHA is not just the regulatory agency it is usually perceived as, but it is also interested in assisting workplaces to improve their safety and health management systems.

The first and foremost of these partnership programs is the Voluntary Protection Programs (VPP). These are a group of three programs that were created in 1982 to provide recognition to those work sites that have demonstrated to OSHA their commitment to provide exemplary safe and healthful working conditions to all of their employees and other workers at their facilities. The VPP also emphasizes the importance of systematic approaches to workplace safety and health based on comprehensive safety and health management systems.

AN OVERVIEW OF THE VPP

The VPP was established by OSHA to implement at a national level a program that had its origins in California in the late 1970s during the construction of the San Onofre Power Plant. The construction of the power plant was under the direction of Bechtel, a company that already had a strong safety and health program that included management leadership and commitment and encouraged employee involvement. Bechtel, the California Building Trades Council, and the National Constructors Association worked together to initiate a joint labor–management safety and health committee to oversee the safety and health activities at the construction project. The committee was responsible for performing routine work-site inspections and the investigation of worker complaints. California OSHA (Cal/OSHA) agreed to empower this committee to perform routine workplace inspections and to not perform any programmed compliance inspections. This program was approved by the California OSHA State Plan and submitted to Federal OSHA for its concurrence, and OSHA agreed to allow Cal/OSHA to proceed with the experiment. At the conclusion of the project, the experiment was deemed a success based on the sense of ownership of the safety and health program expressed by the trades' workers, as well as being one of the safest such construction projects at the time.

After becoming president, former California Governor Ronald Reagan appointed Thorne Auchter as the Assistant Secretary of Labor for OSHA. Being aware of the dramatic success of the San Onofre program, Auchter directed the agency to develop a similar program for OSHA. The primary difference between the OSHA program and that offered by Cal/OSHA to the San Onofre project was that OSHA retained the right to perform inspections related to formal employee complaints and workplace fatalities and catastrophes. VPP work sites were to become exempt from routine programmed OSHA inspections.

To initiate such a program that emphasizes the cooperation between labor, management, and OSHA, the agency had to overcome several obstacles. Traditionally, unions distrusted management when it came to safety and health in the workplace. Second, management typically did not invite OSHA into a workplace to observe the recordkeeping, conditions, and activities at the workplace. This idea harkens to the adage that the biggest lie in workplace safety is that when OSHA arrives at a workplace, it knocks on the door and says: "I'm here from OSHA and I came to help." The second lie is that management then says: "We are glad to see you; we were just about to call for your help." A third obstacle is the fact that OSHA had no formal requirement for safety and health management systems and did not have any guidelines for these systems.

Thanks to the efforts of several OSHA staff and those that responded to the public notice published in the *Federal Register* about the new program named the Voluntary Protection Programs, OSHA announced the VPP in the *Federal Register* on July 2, 1982, to establish the credibility of cooperative action among government, industry, and labor to address worker safety issues and expand worker protection. The success of the VPP is in large part due to those that created and nurtured the program in its early years. Foremost among those OSHA staff was Margaret "Peggy" R. Richardson, who has been dubbed the "Mother of the VPP." The authors wish to recognize Peggy for her dedication to this program and also for her book.[4]

When the VPP was first announced, there were three programs: Star, Try, and Praise. Star was the program designation that was, and still is, assigned to those workplaces that exemplified what OSHA had envisioned as a demonstration of a strong management commitment for worker safety and health and a high level of involvement of employees in safety and health activities. Try was the precursor of the current Merit Program. The purpose of the Try Program was to provide recognition to those workplaces that did not meet the high level of safety and health quality of the Star Program but that were nonetheless committed to improving their management systems for safety and health to the Star Program level. Praise was intended as a program for low-hazard industries.

The very first VPP sites were three Johnson and Johnson Ortho-Clinical Diagnostics sites in New York, Massachusetts, and New Jersey. These were the only such work sites to be approved for the Praise Program. On October 26, 1982, OSHA approved the first workplace into the Star Program, which was General Electric Combustion Engineering in Wellsville, New York. Although ownership has changed several times over the years, that work site, now Alstom Air Preheater Company, is still a VPP Star work site.

In 1985 a group of current VPP sites formed a volunteer association to take over from OSHA the responsibility of managing the annual VPP conferences for VPP sites and to provide potential applicants with pertinent information about the program and to assist OSHA in helping to grow the VPP. With OSHA's assistance, the conferences continued to grow with representation from the core VPP companies as well as those companies that had heard about this new VPP and were interested in learning more about it. In 1990 this group of volunteers decided to form a more formal organization and became the Voluntary Protection Programs Participants' Association, Inc. (VPPPA) in 1991. Peggy Richardson retired from OSHA in 1991 and became the first Executive Director of the VPPPA. This new organization changed its primary function from organizing the annual conferences to "be a leader in health and safety excellence through cooperation among communities, workers, industries and governments."[5] The role of the VPPPA has evolved since its inception as a conference planner to a strong advocate for workplace safety and health and the VPP. It represents over 1900 federal and state plan VPP sites and has assisted OSHA in the review of several standards. The annual conference has also grown from a small gathering of about 50 representatives to major safety and health conferences with over 3000 attendees. The VPPPA has also expanded its availability through the organization of regional chapters in each of the 10 OSHA regions. These regional chapters also hold, albeit smaller, annual conferences. In addition to offering workshops directly related to the VPP, each of these 11 conferences offer workshops on specific safety and health elements to help attendees improve their safety and health management systems and to network with others with the same goals.

EVOLUTION OF THE VPP

Since its inception, OSHA has regularly reviewed the progress and effectiveness of the VPP and has made some modifications to it. On October 29, 1985,[6] OSHA issued its first revision to the VPP. The most significant change was to no longer allow sites to qualify based solely on their safety program; sites now had to provide effective protection against both safety and health hazards. Based on the determination that since there were no additional applications for Praise, there was little reason to continue the program and the Praise Program was discontinued. The revisions also established the VPP Demonstration Program to replace the experimental element of the Try Program. The purpose of the Demonstration Program is to "provide the opportunity for companies to demonstrate the effectiveness of alternative methods, which if proved successful (usually at more than one site), could be substituted for the Star Program for certain situations." Another purpose of the Demonstration Program was to "test methods of overcoming problems which have kept certain employers, such as small business employers and many contractors in the construction industry, from taking part in the VPP."[7] The Demonstration Program eventually began to be referred to as the Star Demonstration Program since the requirements to participate were intended to be as stringent as those for the Star Program.

In January, 1988, the Try Program was officially changed to what is now the Merit Program. The intent of the Merit Program is to recognize employers in any industry who do not yet meet the requirements for the Star Program but who have demonstrated the commitment and potential to achieve Star requirements within an agreed period of time that may not exceed 3 years. Although the VPP has undergone a few more revisions, it remains much the same as it was in 1988.

The more significant changes were to establish several demonstration programs to address the transition by the Bureau of Labor Statistics from the use of the Standard Industrial Classification (SIC) codes to the newer North American Industrial Classification System (NAICS).

It is interesting to note that when the VPP was initiated OSHA had no formal standards or guidelines for safety and health management systems. It was not until January, 26, 1989, that OSHA published its "Safety and Health Program Management Guidelines; Issuance of Voluntary Guidelines" in the *Federal Register*. It is also interesting to note that the guidelines were based on the positive experiences of the VPP sites. There were documented reports of significant reductions in injuries and illnesses and related direct and indirect cost savings. Once these guidelines were published, it led to another change to the VPP to bring the VPP's basic program elements into conformity with OSHA's Safety and Health Program Management Guidelines of January 26, 1989. Another change was to formally include resident contractors at participating VPP sites as potential applicants for their activities at those VPP sites. Other subsequent revisions were more procedural, such as the revision to recognize the changes to the OSHA log and the initiation of the use of TCIR (total case incident rate) and DART (days away, restricted, transferred cases). With the 1988 and later revisions, the official programs became the current versions of Star, Merit, and Demonstration.

The next significant revision to the VPP occurred in July, 2000, when OSHA added a new method to calculate rates for small businesses that made it easier for them to meet the Star rate requirements. Instead of using just the most recent three calendar years on injury/illness rates, small businesses that meet a specific criteria are now allowed to consider the best three out of four year's injury/illness rates. The rational for that decision was that in a company with only a few employees, any one or two recordable injuries could remove them from eligibility for the VPP.

Another revision to the VPP in July, 2000, was the formal inclusion of recordable illnesses in the rate calculations. Until then only injuries were included in the calculations.

On January 9, 2009, OSHA published in the *Federal Register* the latest changes to the VPP. These changes become effective on May 9, 2009. The most significant of the changes include:

- Acceptance of the Corporate VPP Pilot Program as a formal part of the VPP
- Acceptance of the Mobile Workforce for Construction Demonstration Program as a formal part of the VPP
- Modified provisions concerning Star Program rate reduction plans and 1-year conditional status
- Greater emphasis on the principle of continuous improvement
- Formal expectation of outreach and mentoring activities

VPP FOR FEDERAL AGENCIES

Another significant date in the VPP chronology was October 27, 1997. That was the date that OSHA extended the eligibility for the VPP to federal agencies. Since the OSH Act does not provide OSHA with formal jurisdiction over other federal agencies, this was a very significant VPP milestone. Several federal agencies, including the U.S. Park Service, the Department of Defense, and the U.S. Postal Service (USPS) have become VPP sites. In fact, the USPS considers the VPP such a value-added safety and health tool that it has fully embraced the program. Since 1997 the USPS has successfully had over 150 individual sites approved into the VPP. However, this was not the first federal incursion into the VPP. In 1994 the U.S. Department of Energy initiated its own VPP, which was based in a large part on OSHA's VPP.

The Department of Defense has, for several years, recognized the importance of the VPP as a valuable tool to maintain a strong focus on the safety and health of its workers. Since OSHA has opened the doors to federal agencies to apply to the VPP, the Department of Defense has committed to identifying and working with its leading installations toward VPP participation. Several military establishments are currently in the VPP and more are working toward the goal of participation.

Even OSHA has gotten on the VPP bandwagon with seven of its area offices and other specialty operations recognized as Star work sites.

VPP STAR PROGRAM

The OSHA VPP Star Program is the highest level of recognition for safety and health management systems offered in the United States. The Star Program is designed for VPP sites whose safety and health management systems operate in a highly effective, self-sufficient manner and meet all VPP requirements. As the highest level of VPP participation, Star represents those sites that exemplify the strongest level of commitment to safety and health in the workplace. To qualify for Star, all safety and health management system elements and subelements must meet or exceed the expectation of OSHA and be determined to be effective for at least the most recent 12 months.

In addition, the rates for recordable injuries and illnesses for the most recent 3 full calendar years prior to the submission of an application must be lower than the most recent average rates published by the Bureau of Labor Statistics (BLS) for the specific industry classification. Based on a recent change to the VPP criteria, OSHA allows the comparison to be based on any of the most recent 3 years of BLS published rates. This decision was a result of the realization that it was unfair to compare a company's 3-year rates to a single year average rate as published by the BLS.

Once approved for Star, there is no official termination period. Participation continues uninterrupted based on routine follow-up evaluations by OSHA. After the initial approval to Star, OSHA reevaluates the work site within the next 36–42 months. Upon satisfactory findings of that evaluation, the site is allowed to continue and will be reevaluated again within 60 months from the completion of each subsequent OSHA evaluation.

The routine evaluations are supplemented with annual status reports submitted to OSHA by each VPP work site. Each year, all VPP sites must provide to OSHA, by February 15, a report that includes a table that contains the OSHA recordkeeping log data for the past 4 years and a copy of the annual evaluation for the previous year. That provides OSHA a means to maintain a constant review of the site's continued meeting of the requirements of the VPP. Should OSHA perceive that there has been a deterioration of the quality of the safety and health management system, it may revisit the site for an interim evaluation and place the site on a 1-year conditional approval.

A 1-year conditional approval allows the site to work on specific 1-year conditional goals to bring the quality of its safety and health management system back to Star quality.

These goals are issued to address weaknesses such as failure to perform monthly inspections, failure to track hazard correction items, missed training programs, and weak inspections caused by a lack of hazard recognition training. After one year, OSHA revisits the site to evaluate the progress made in completing the goals. Successful completion may result in the site being reinstated in the Star Program. Failure to meet any goal will usually result in a formal request by OSHA that the site withdraw from the VPP without prejudice. After the withdrawal, the site may continue to work on the goals and then submit a new VPP application in the future.

The 1-year conditional goals are not used to directly address increases in injury and illness recordable rates that cause the BLS average rates to be exceeded. Should the site realize an increase in its rates, or should the BLS rates drop to a level below the site's rates, OSHA will issue a 2-year rate reduction letter. The site will be given 90 days to develop a plan to address those factors that resulted in the rate increases. When the rate reduction plan is approved by OSHA, the site will have 2 years to implement it with the goal of bringing the rates to below those of BLS and continue as a Star site. It must be noted that the goal must be specific and action based. A goal such as "We will reduce our rates to below the BLS rates within two years" is not acceptable except as a result of successful completion of the specific actions.

Examples of rate reduction goals include: provide hazard recognition training to workplace inspectors; ensure through tracking that inspections are performed at least monthly; strictly enforce all site safety and health rules such as lockout/tagout; and develop and implement a new recognition program for employees to encourage their participation in safety and health activities.

VPP MERIT PROGRAM

The Merit Program provides recognition to applicants that have good safety and health management systems but that are either not yet at the Star level of quality or they have not been fully in place and effective for at least 12 months. It is possible to be approved for Merit if all of the elements are at least operational or, at a minimum, in place and ready for implementation by the date of approval.

In addition, all minimum VPP requirements must be met. The minimum VPP requirements are also applicable to the Star and Star Demonstration programs and include:

A. Management Leadership and Employee Involvement
   1. A written safety and health management system at least minimally effective to address the scope and complexity of the hazards at the site.
   2. Management demonstrates at least minimally effective, visible leadership with respect to the safety and health program.
   3. Top management accepts ultimate responsibility for safety and health in the organization.
   4. The individuals assigned responsibility for safety and health have the authority to ensure that hazards are corrected or necessary changes to the safety and health management system are made.
   5. Adequate resources (equipment, budget, or experts) are dedicated to ensuring workplace safety and health.
   6. The site's contractor program covers the prompt correction and control of hazards in the event that the contractor fails to correct or control such hazards.
   7. Contract oversight is minimally effective for the nature of the site (inadequate oversight may be indicated by significant hazards created by the contractor, employees exposed to hazards, or a lack of host audits).
   8. Employees support the site's participation in the VPP process.
   9. Employees feel free to participate in the safety and health management system without fear of discrimination or reprisal.

B. Work-Site Analysis
   1. The site has been at least minimally effective at identifying and documenting the common safety and health hazards associated with the site (such as those found in OSHA regulations, building standards, and the like, and for which existing controls are well known).
   2. There is at least a minimally effective hazard analysis system in place for routine operations and activities.

3. The site has a minimally effective system for performing safety and health inspections and identifying hazards associated with normal operations.
4. There is a minimally effective means for employees to report hazards and have them addressed.
5. A minimally effective tracking system exists that results in hazards being controlled.
6. There is a minimally effective system for conducting accident/incident investigations, including near misses.
7. The site has a minimally effective means for identifying and assessing trends.

C. Hazard Prevention and Control
1. The site selects at least minimally effective controls to prevent exposing employees to hazards.
2. The site has minimally effective written procedures for emergencies.

D. Safety and Health Training
1. The site provides minimally effective training to educate employees regarding the known hazards of the site and their controls.

OSHA may also award Merit recognition to those applicants that do meet the Star requirements with the exception of the rate requirements. If it is determined by the evaluation team that an applicant has demonstrated the commitment and possesses the resources to meet Star requirements within 3 years, the employer may enter the Merit Program with set goals for reaching Star. If the rates represent an issue, then the applicant must be able to demonstrate that it is programmatically and statistically feasible to reduce rates to below the industry average within 2 years. If the applicant has either or both the TCIR and DART rate above the industry average, the applicant must set realistic, concrete goals for reducing both rates within 2 years and must specify the methods (approved by the VPP Manager) to be used to accomplish the goals. A Merit applicant would qualify for Star when it has met its Merit goals, the Star rate requirements, and when all other safety and health elements and subelements have been operating at Star quality for at least 12 months.

When an applicant is approved as a Merit site, it is assigned goals that must be met within 3 years to be able to maintain continued participation until it qualifies for the Star Program. The Merit goals address Star requirements not in place during the initial evaluation or aspects of the safety and health management system that are not up to Star quality for at least 12 months. The Merit goals include methods for improving the safety and health management system to address the identified problem areas. Merit goals may also address weaknesses in safety and health management system deficiencies underlying the high recordable rates with the intent of reducing a 3-year TCIR or DART rate to below the national average. A goal to reduce the rates would not in itself be a valid Merit goal.

Following are some examples of Merit goals:

1. Improve the accident investigation process to include more extensive training for those performing the investigations and ensure more appropriate corrective

actions, with a focus on revisions to procedures, equipment, and training. Success of this Merit goal will be demonstrated by a review of future accident investigations by a VPP evaluation team.

2. Employees must be encouraged to participate in all aspects of the safety and health management system. This may include their participation in more frequent inspections, providing training, helping to perform hazard assessments, making formal suggestions, and being involved in production and process modifications. Success of this Merit goal will be demonstrated by employee interviews and reviews of their activities.
3. Training records and programs must be improved to better document the scope of the training and should be provided with greater frequency. Success of this Merit goal will be demonstrated by reviews of all training records and schedules.
4. The annual evaluation should be better detailed to include information as to how observations and recommendations were arrived at. It must also include a method to ensure that all actions on previous recommendations have been completed. Success of this Merit goal will be demonstrated by a review of the next annual evaluation.
5. Clearly define the responsibilities of the management employees in relation to safety and health. The performance appraisal should include specific areas of accountability for each rated employee relative to his or her safety and health responsibilities. Success of this Merit goal will be demonstrated by reviews of new performance evaluations and interviews with selected managers.
6. Reduce the rates to below the industry average within 2 years by the following activities:
    a. Develop a plan for determining the factors contributing to the elevated numbers including identifying any injury/illness trends.
    b. Develop a plan for reducing the number of injuries/illnesses necessary to reduce the 3-year rate to below the most current BLS averages.
    c. Develop and implement specific objectives for accomplishing the rate reduction plan.
    d. Success of this Merit goal will be demonstrated by a review of the OSHA logs and the VPP rates table.

Both Star and Merit applicants and current participants may also receive a list of what OSHA refers to as 90-day items. These include those compliance-related issues and workplace hazards that were observed during the VPP evaluation team's work-site tour and program revisions that were not corrected by the conclusion of the evaluation. However, when a safety and health management system deficiency underlies a specific hazardous condition, then corrections to the system must be included as Merit goals.

The length of a Merit term for approval is dependent on the estimated time necessary to fully accomplish the Merit goals. However, initial approval to Merit will be for a single term not to exceed 3 years. In exceptional situations the OSHA assistant secretary may allow an additional 3-year extension to the Merit term.

One example of a reason for such an extension may be a significant interruption of the work process or damage to the plant that prevented the completion of the Merit goals. The 2005 devastation caused by Hurricane Katrina in New Orleans and Mississippi would probably represent a valid disruption and a reason for an extension to the Merit term.

GROWTH OF THE VPP

For the first 10 years of the VPP, there was a steady but slow growth. The only work sites that were eligible were those covered by federal OSHA since no state plan state had yet started its own VPP. Also, there was no strong effort to encourage participation in the VPP, and the growth was primarily as a result of word of mouth and a commitment by a few major companies such as Mobil Chemical, which by July, 1987, had all of its 24 work sites in the Star Program, and the General Electric Company, which began slowly but eventually became the foremost corporate presence in the VPP with over 100 VPP work sites in 2008. In fact, General Electric has developed its own internal VPP-type program called the GE Global Star. Recognizing the benefits of the VPP as a management system for safe and healthful workplaces, the GE Global Star offers corporate recognition to those international General Electric facilities that have met the equivalent of the VPP criteria. Other major corporations that have adopted the VPP as a management tool to improve workplace safety and health include Monsanto, Milliken, Georgia Pacific, Covanta Energy, and International Paper.

With the new emphasis on safety and health management systems and the focus on the VPP, OSHA made an all-out effort to encourage corporations and individual companies to participate. The VPP saw a growth from the 100 sites in 1992 to 250 sites in 1996, for a growth of 150% in 4 years. Clearly, the word was getting out about this unique and very successful program. The experiment to demonstrate that labor, management, and OSHA can effectively work together to prevent workplace injuries and illnesses and reduce costs was working. The VPP was not considered the "flavor of the month" safety program but rather as a new way of managing safety as an ongoing commitment.

Federal OSHA has jurisdiction in 28 states, plus the Virgin Islands. The other 22 states, plus Guam and Puerto Rico, are responsible for workplace safety and health through their state plan agreements with federal OSHA. Through those state plan agreements, the states run their own safety and health organizations and receive funding for 50% of their costs from federal OSHA. State plan states may promulgate their own workplace safety and health regulations. However, the primary requirement of the state plan programs is that they must be at least as effective as federal OSHA in their workplace safety and health rules. For example, their programs for fall protection must be at least equal to OSHA's, but they may exceed the OSHA rules. Where OSHA requires fall protection in construction starting at elevations of 6 feet, state plans may require such protection at 4 feet. This is applicable to the

VPP as well. In developing their VPP, the state plan states or territories may decide to use the OSHA model in its entirety, or they may make it more restrictive. Whereas the federal OSHA staffs are federal employees, state plan OSHA staffs are state employees.

In August, 1986, California became the first state plan state to start a VPP. That is only fitting given that the OSHA VPP was originally based on the Bechtel experiment in San Onofre, California, 7 years earlier. The other states followed with their own VPP models with Vermont becoming the last state plan state to offer the VPP to sites under its jurisdiction. In accordance with the OSH Act, state plan states must be able to demonstrate that their programs are at least as effective as those similar programs of OSHA. Many of the state plan states have made the requirements of their VPP even more stringent than OSHA's. For example, the South Carolina VPP does not have a Merit Program and requires its Palmetto Star Program sites to have injury/illness rates 50% below the state averages for similar industries.

Probably the most unique site to raise the VPP Star flag was Task Force Rakkasan, Kandahar, Afghanistan, made up of units from the U.S. Army, Air Force, Marine Corps, and a Canadian battle group. Although the military is not eligible to participate in the VPP, an honorary flag was presented to that base to recognize the efforts made to meet the requirements of the VPP. Credit must be given to Dave Baker, formerly the Regional VPP Manager for OSHA Region 10, and his fellow members of Task Force Rakkasan.

Since those early spurts of minimal growth, the VPP received a major emphasis by OSHA and the Congress of the Clinton White House. It was not until 1992 that the VPP reached 100 sites.

The next milestone was the recognition of the 500th federal OSHA VPP site in February, 2000. That was awarded to a resident contractor at the NASA Johnson

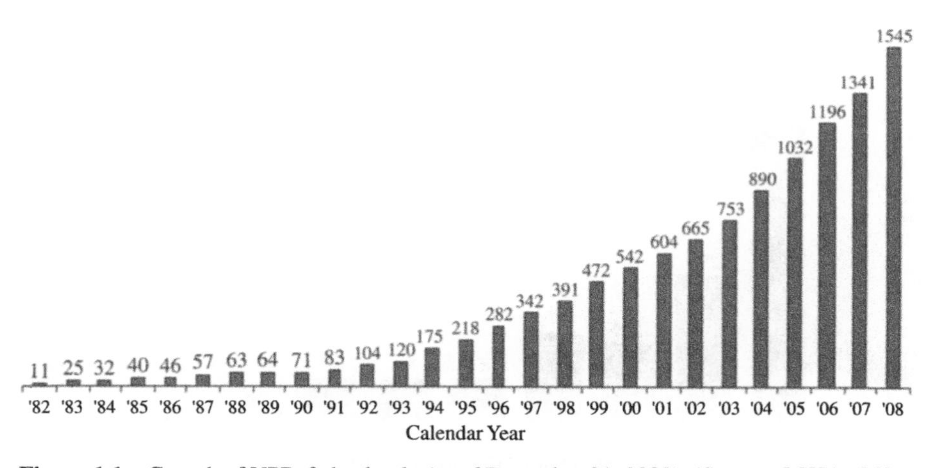

Figure 1.1 Growth of VPP; federal only (as of December 31, 2008). (*Source*: OSHA, Office of Partnership and Recognition.)

Space Center in Houston, Texas, under the Demonstration Program for Resident Contractors. That represented a 200% increase in another 4 years. It was only another 3 years to the 1000th site, the Titleist Ball Plant II in Dartmouth, Massachusetts, in October, 2003. The growth of VPP continued to July, 2008, when OSHA issued the VPP Star flag to Wyeth Pharmaceuticals, Pearl River, New York, to make it the 2000th VPP site. As of December 31, 2008, there were over 2149 VPP sites in all of the 50 states. Figure 1.1 shows the progressive growth of the VPP from its inception in 1982 to November 30, 2008. This figure includes all VPP sites in all states, both federal and state plan programs. It also includes all federal agency sites recognized as VPP sites.

VPP MEMBERSHIP

The VPP has no restrictions for participation based on either the size of the work site or the type of industry. Figure 1.2 illustrates the participation in the VPP by work-site size and industry.

There are no restrictions to participation in the VPP. So long as either federal or state OSHA has jurisdiction over a workplace, that workplace may apply for the VPP. Workplaces may apply regardless of how many employees work there. One of the smallest VPP sites has only 19 employees and the largest has had over 4000. There is also no restriction for sites with or without union-represented workers. There is one additional requirement for nonconstruction workplaces that have union-represented employees working directly for the company. At those workplaces the senior union official at the workplace must not oppose the VPP application. Evidence of that must be included in the VPP application. That evidence may take the form of a concurring signature on the application indicating support for the VPP

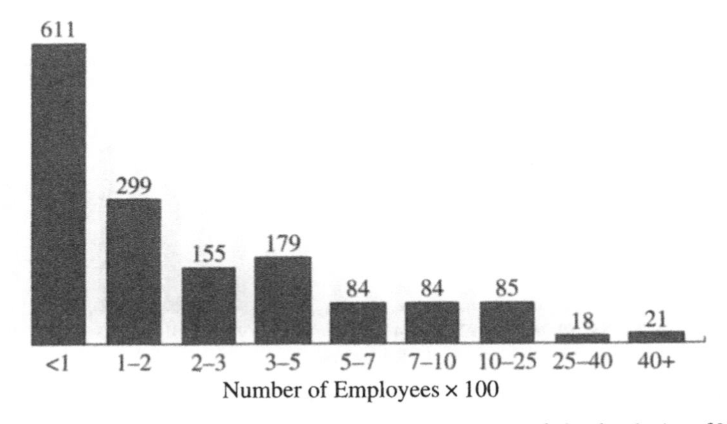

Figure 1.2 Size of VPP sites. Number of sites by employment; federal only (as of November 30, 2008). (*Source*: OSHA, Office of Partnership and Recognition.)

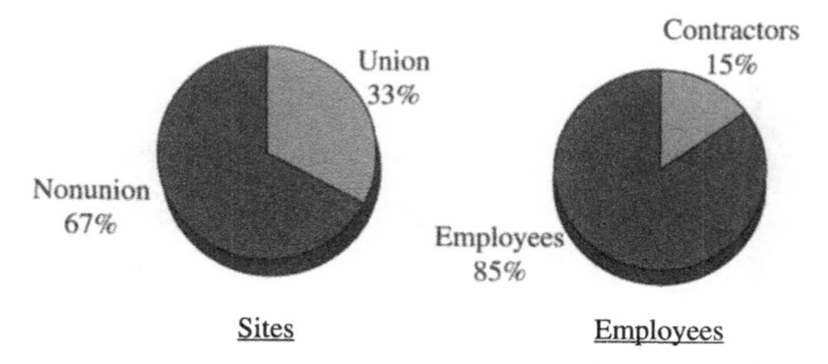

Figure 1.3 Union and nonunion VPP sites; federal only (as of December 31, 2008). (*Source*: OSHA, Office of Partnership and Recognition.)

or a statement that the union has no objection to the VPP application. Clearly, the former alternative is preferred since it indicates for the record that the union is in favor of the VPP. Figure 1.3 illustrates the percentage of union versus nonunion VPP work sites and the respective numbers of employees in each category.

Construction projects that hire only union workers must also obtain a statement of support or nonobjection to the VPP from each of the represented unions. That is usually obtained from a group that represents all local trades unions, such as a construction and building trades association. Those construction work sites that hire both union and nonunion workers should use the criteria detailed in Table 1.1 to determine if they must obtain official union support for the VPP application.

The chemical industry includes chemical companies and oil refineries, and the trucking and warehousing industry includes the U.S. Postal Service. The electric industry consists of sites that are involved in any one or more of electric generation, transmission, or distribution, including a large number of resource recovery sites. Figure 1.4 illustrates the industries with the most VPP work sites. It indicates that there are currently 28 general building contractors in the VPP. That number does not reflect all of the construction projects that have been in the VPP because when the project is completed the VPP recognition also ends for that project.

TABLE 1.1 Criteria to Determine If Construction Work Sites Must Obtain Official Union Support for the VPP Application

If	Then
Majority of employees are represented by unions	Signed statement(s) required. Must be obtained from enough unions to represent a majority of employees.
Some employees but less than a majority are represented by unions	No statement of union support required.
No employees are represented by unions	Requirement not applicable.

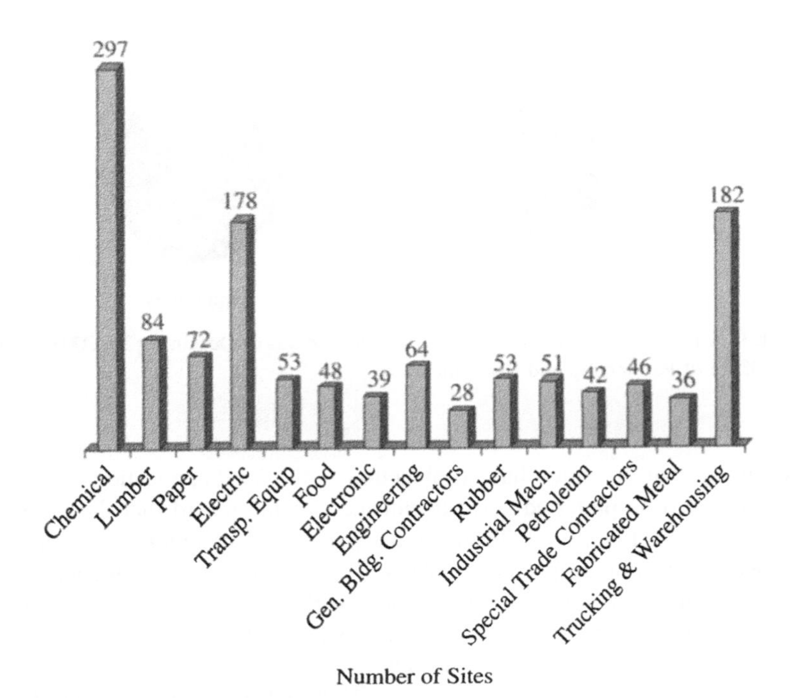

Figure 1.4 Top 15 industries in the VPP; federal only (as of November 30, 2008). (*Source*: OSHA, Office of Partnerships and Recognition.)

CONSTRUCTION INDUSTRY IN THE VPP

The standard requirements of the VPP for general industry preclude most construction projects from participating in the VPP. In recognition of that issue, OSHA has included the following exceptions in the VPP Policies and Procedures Manual[8] exclusively for the construction industry. These exceptions allow larger construction projects to apply to the conventional VPP. To be able to apply, a construction applicant must be the general contractor, owner, or an organization that provides overall management at a work site, controls site operations, and has ultimate responsibility for assuring safe and healthful working conditions at the work site. Instead of the 3 years experience required of general industry applicants, construction projects must have been in operation for at least 12 months prior to approval. The rate calculation is also reduced from 3 years to just the amount of time that the project has been in operation when the application is submitted. If the project's TCIR and DART rates are below those for the general construction BLS rates, the project may qualify for Star. If the project's rates are above the BLS rates, the applicant may still qualify for Merit if the company-wide 3-year TCIR and DART rates are below the national average.

Unlike the general industry VPP, construction VPP projects must submit an end-of-project final comprehensive evaluation. To address the continually changing conditions of construction projects, inspections must be conducted at least weekly with the entire project inspected at least monthly. Since the VPP covers all workers at the project, all hazards must be corrected, including those created by subcontractors.

Similar to a conventional VPP application, all employees must be made aware of the VPP application or participation and of their rights, roles, and responsibilities. That includes the employees of the applicant as well as those of all subcontractors. Evidence that all subcontractors at the work site recognize these conditions is necessary and may include:

1. The contractual agreement
2. A written statement of willingness to cooperate
3. Attendance at safety meetings
4. Orientation sessions for incoming subcontractor employees

Employees at construction sites must be involved in safety and health at the work site to the degree practical based on the time they will spend on site. Examples of short-term involvement include attending daily toolbox talks on safety and health and participating in daily self-inspections. The more time they spend on site, the more involvement OSHA expects. The onsite evaluation team will judge the sufficiency of employee involvement through interviews and observations.

OSHA SPECIAL GOVERNMENT EMPLOYEES PROGRAM

OSHA has implemented several initiatives to improve the VPP. On February 28, 1994, it initiated the OSHA VPP Special Government Employee (SGE) Program, and in June, 1994, it formalized the VPP Mentoring Program. The SGE Program was established to allow industry employees from VPP work sites or companies to work alongside OSHA during VPP onsite evaluations. One of the benefits OSHA realizes from this program is that this helps to supplement the OSHA onsite evaluation teams with other than OSHA staff, thereby allowing those OSHA resources to be used in other ways to meet the agency's goals. Another benefit for OSHA and the VPP is the availability of subject matter experts participating on VPP evaluations. Those subject matter experts have included chemical safety, cranes, and ergonomics.

The SGE Program has an added benefit in that the SGEs receive another opportunity to improve their safety and health and VPP process. It provides the SGEs exposure to how other companies manage their safety and health programs. Experienced SGEs have confirmed that they have always been able to bring something back from these exposures to their own work sites to improve their safety and health management system.

To become an SGE an applicant must meet the following requirements:

- Employee of a current VPP site or a corporate employee of a corporation with VPP sites
- Strong interpersonal skills

- Sound reading and writing skills
- Physical ability to perform team member's duties
- Management or corporate support for participating as an SGE experience applying OSHA regulations
- Experience (currently or within the previous 2 years) in a leadership position(s) in the VPP at the applicant's work site or corporation (this includes hourly employees directly involved in the VPP process regardless of their safety and health experience or education)

The interested employee must complete and submit a detailed application to OSHA. The application asks for the appliant's safety and health and VPP experience and a background of the applicant. It also includes a full financial disclosure to assist OSHA in confirming that there is no potential conflict of interest with sites being evaluated. Once the application is reviewed and accepted by OSHA's national office, the applicant is assigned to an SGE training class. There are at least four SGE classes throughout the country each year. These are supplemented with other classes on an as-needed basis. Many VPP companies have sponsored SGE classes so that their own applicants can be trained locally.

The 24-hour SGE training course reviews the history of the VPP, the elements of safety and health management systems, and the techniques to evaluate those elements. The training also explains the government code of ethics that apply to all federal employees, including the SGEs. At the conclusion of the SGE training, all participants are formally sworn in by an OSHA official, usually the local Area Director.

The SGEs then volunteer to assist on OSHA VPP onsite evaluation teams, given any company restrictions for travel and expenses. As a team member, SGEs are held to the same standards as regular OSHA employees. The only exception is that only OSHA employees may review the OSHA logs and supporting personal medical records.

In addition to supporting OSHA in its VPP activities, many SGEs assisted OSHA in its activities during the World Trade Center Recovery Program. SGEs from companies across the country helped OSHA distribute personal protective equipment (PPE) to the recovery workers. That included the quantitative respirator fit testing of over 4000 members of the Fire Department of New York and the distribution of over 130,000 respirators and other PPE.

In December, 2004, OSHA began to allow SGEs to waive the cost for courses offered at the OSHA Training Institute (OTI). These courses are the same courses offered to the OSHA compliance safety and health officers and staff.

VPP MENTORING PROGRAM

The VPP Mentoring Program was initially established by OSHA to provide a means for current VPP sites to assist or mentor prospective VPP applicants in the application process. When it was first set up by OSHA, a prospective applicant or mentee would contact the regional VPP manager or the OSHA national office to recommend a mentor to establish an informal match. This program was transferred to the VPP Participants' Association and renamed the VPPPA Mentoring Program.

The function of the mentor is not specifically defined and the actual assistance provided by the mentor is decided upon by the mentor and mentee. At a minimum, the mentor provides assistance to the mentee to address the administrative requirements of the VPP and the application process. The administrative requirements include such matters as notifying all employees about the VPP and the intent to submit an application, the methods to review the various safety and health programs, the techniques to prepare the site for the OSHA onsite evaluation, and other related issues. The application assistance may include suggestions about the actual writing of the application as well as what the OSHA VPP manager expects in an application. The mentor does not usually get involved in the writing or the review of hazard control programs or an inspection of the mentee's work site other than for familiarization. Through programs such as the SGE Program and the VPPPA Mentoring Program, the VPP has been able to continue its growth.

DEMONSTRATION PROGRAMS

John Henshaw, the former Assistant Secretary of OSHA, in September, 2002, announced that the goal of the agency would be to increase participation in the VPP from 800 sites to 8000 sites. Although that was a stretch goal, it set the agenda for the VPP for the following years. To achieve that goal, OSHA created several unique new demonstration and pilot programs. However, these were not the first demonstration programs. The first such program was initiated in 1992 to allow resident contractors at current VPP sites to apply on their own merits for the work performed at the VPP site. The first such approval was made in early 1993. This demonstration program existed until 2000 when it was approved as another formal eligibility category for VPP applicants.

DEMONSTRATION PROGRAM FOR SHORT-TERM CONSTRUCTION PROJECTS

The next demonstration program was initiated in April, 1998, to allow short-term construction projects to apply to the VPP. OSHA recognized that the VPP requirements for construction projects were very restrictive since the application could not be submitted until the construction project was in operation for at least 12 months. Allowing for another 6 months for the application process to proceed to approval, most construction projects would not be able to apply. There are very few construction projects that extend beyond 18 months to 2 years. Hence, the most hazardous industry was generally precluded from participation in a program that would enhance safety and health at construction projects.

The Demonstration Program for Short-Term Construction Projects was established to overcome the obstacles by allowing construction general and trades contractors to apply to the VPP for their shorter-term projects. This was accomplished by having construction companies submit an application based on the company's corporate programs and injury/illness records. OSHA would then evaluate the company's corporate structure including a review of the organization of safety and health in the corporate structure, assignments and responsibilities for managers and supervisors, resources for safety and health activities, all hazard control programs, procedures for worksite analyses, and the level of management leadership and commitment and employee involvement. When the corporate structure had been approved, the company could then submit an abbreviated application for individual projects that were still in the beginning phases of work. If the individual projects met the requirement, they would be approved as VPP sites under the demonstration program.

MOBILE WORKFORCE DEMONSTRATION PROGRAM

The Mobile Workforce Demonstration Program was initiated in late 1998. This program was for nonconstruction companies with mobile workforces to demonstrate that they could effectively protect their field employees regardless of their work locations. Examples of these mobile workforces include appliance repair services and field power generation maintenance workers such as linemen. This program was similar to the Construction Short-Term Demonstration Program.

3-C DEMONSTRATION PROGRAMS

In 2003 OSHA initiated three new demonstration programs that were developed to respond to Assistant Secretary John Henshaw's goals to increase the coverage and effectiveness of the VPP. These became known as the 3-C demonstration programs.

MOBILE WORKFORCE FOR CONSTRUCTION DEMONSTRATION PROGRAM

The first 3-C program was the Mobile Workforce for Construction Demonstration Program. This program was developed by combining the former Demonstration Program for Short-Term Construction Projects and the Mobile Workforce Demonstration Program. Although the primary industry focus of this demonstration program is the construction industry, all types of industry classifications may apply.

This demonstration program is intended to create greater opportunity for employers and employees in the construction and other industries to participate in the VPP and, in so doing, to strengthen worker protections significantly. At the same time, it is intended to provide OSHA with additional opportunities to explore and test appropriate modifications to the VPP and the administration of alternate safety and health management systems. These alternative requirements should help OSHA bring the benefits of this program to the construction industry and other underrepresented industries.

The first step in the application process for this demonstration program is to create a dialog with the OSHA Regional VPP Manager. The Regional VPP Manager will discuss the requirements of the program and geographic extent of the application. The geographic coverage of the application is referred to as the designated geographic area (DGA), which must be agreed to by OSHA and the company. The usual DGA is that area covered by a single OSHA area office. The DGA may be extended to a larger area based on the size of the area and the number of active applicant work sites in the DGA. Although it is possible to extend the DGA across OSHA regional jurisdictions, that is very rare and must be arranged with the cooperation of all involved OSHA regions.

The applicant company completes a special application that details the corporate or company safety and health management system and the corporate or company structure for safety and health. It also details how the company provides for the safety and health of its workers at their remote work sites. The application includes a description of the type of work and the types of locations at which the work is done. The application process also requires a participation plan, which that focuses on the following information:

1. Designated geographic area (DGA)
2. Unique aspects of company's mobile workforce
3. Subcontractor oversight
4. Hazard recognition and control as a noncontrolling employer
5. Employee involvement
6. Baseline hazard analysis
7. Emergency response

Once the application has been accepted, OSHA will start the evaluation phase. Unlike a typical evaluation, the evaluation for this demonstration program is actually a multiphased process. OSHA will first visit the company headquarters to evaluate the company records, including the injury and illness information and specific safety and health hazard control programs. OSHA will also interview several company managers and employees. The purpose of this visit is to verify the information in the application.

Upon successful completion of the company headquarters evaluation, OSHA will select a small number of active sites within the DGA for evaluation. These evaluations focus on the information in the participation plan and include observations of the work and work site and employee interviews. It is important to note that the work and activities of noncompany workers will also be observed and are expected to be safe with all hazards properly controlled.

Having evaluated this demonstration program and determining that it is an effective method for such companies to participate in the VPP, OSHA has again revised the VPP so that beginning on May 9, 2009, employers with mobile workforces will be able to apply for VPP participation without the need for the demonstration program. This change opens new opportunities for participation by exemplary employers in the construction industry plus mobile workforce employers in other industries.

VPP CORPORATE PILOT PROGRAM

The second of the 3-C demonstration programs is the VPP Corporate Pilot Program. Recognizing that many corporations have committed to having their numerous sites participate in the VPP, OSHA developed this program to try to streamline the application and evaluation process.

The VPP Corporate Pilot Program is designed to test new streamlined VPP processes for corporate applicants, who demonstrate a strong corporate commitment to employee safety and health and the VPP. These applicants, typically large corporations or federal agencies, have either already adopted VPP on a large scale or are in the process of doing so. VPP Corporate Pilot applicants and participants must have established VPP experience for at least some of its facilities, standardized corporate-level safety and health management systems, and hazard control programs, and effectively implemented organization-wide as well as internal audit/screening processes that evaluate their facilities for safety and health performance. They must also have injury/illness rates that compare favorably to the BLS average rates for their primary industry. Under the VPP Corporate Pilot Program, streamlined processes have been established to eliminate the redundancies associated with multiple applications and onsite evaluations, and expand VPP participation for corporate applicants in a more efficient manner.

Criteria that are required to apply to the OSHA VPP Corporate Pilot Program include:

- Significant corporate participation in VPP and a commitment to strengthen VPP participation
- Effective internal prescreening processes
- VPP knowledge and dedicated resources to VPP
- Commitment to outreach and mentoring
- Community leadership in safety and health
- Participation in the Special Government Employee Program

Current participants for the VPP Corporate Pilot are:

- Delta Air Lines
- The Dow Chemical Company
- Fluor Corporation
- General Electric Company
- Georgia-Pacific Corporation
- Parsons Corporation
- U.S. Postal Service
- Washington Division of URS Corporation

Since this is a pilot program, participation is limited. Interested corporations should initially contact the OSHA National Office, Directorate of Cooperative and State Programs, Office of Partnerships and Recognition to express their interest in the Corporate Pilot Program to determine if they qualify and if OSHA can still accept new applications. As with the regular VPP, it is also a good idea just to let OSHA know of the intent to file a VPP application. With OSHA's concurrence, the corporation VPP coordinator will prepare and submit the VPP corporate application describing corporate-level policies and programs consistent with VPP criteria that apply to all facilities across the organization. In addition to meeting the traditional VPP criteria, the applicant must have effective internal prescreening processes to evaluate the candidate facilities' level of preparedness for participation in VPP.

Following OSHA's review and acceptance of the VPP Corporate application, a comprehensive onsite corporate program evaluation at the corporate office/headquarters is conducted by OSHA to verify the information in the application. With the exception of the site tour, the corporate evaluation is similar to a standard site-specific evaluation. It will include interviews with senior leadership and management and safety and health staff to verify their commitment and leadership, interviews with selected managers, supervisors, and hourly employees at both existing corporate VPP sites, and non-VPP sites, evaluation of the prescreening process for prospective applicants, and a comprehensive review of corporate-level policies and programs.

Upon acceptance of the participant into the VPP Corporate Pilot Program, all eligible participant facilities will follow the streamlined application and onsite evaluation process when applying for VPP participation as described below:

- *Corporate–Facility Application Process (C-FAP)* The facility prepares and submits a VPP application using a proscribed format that explains the safety and health management system and includes facility-specific information. Information submitted in the corporate application does not need to be repeated, but the site application must explain any deviations from or additions to the corporate programs.
- *Corporate–Facility Onsite Process (C-FOP)* The facility onsite evaluation focuses on the implementation of the standardized corporate safety and health policies and programs and any facility-specific programs. Also, the duration of the onsite evaluation under C-FOP is shortened using the VPP Corporate Pilot onsite protocol. However, it is the same three-step evaluation consisting of a site tour, employee interviews, and documentation reviews of the OSHA logs and supporting information facility-specific programs. Determinations of how effectively all of the programs have been implemented will be made by the OSHA Team.

Having evaluated this demonstration program and determining that it is an effective method for such companies to participate in the VPP, OSHA has again revised the VPP so that beginning on May 9, 2009, corporations will be able to submit VPP applications under the tested streamlined process. This will enable even more deserving work sites to become members of the elite VPP in an efficient and less resource-intensive manner. The current corporate VPP companies will be moved into the appropriate VPP divisions. The success of both the construction and corporate demonstration programs will enable the VPP to continue to grow in importance in the United States.

OSHA CHALLENGE PROGRAM

The third of the 3-C demonstration programs is the OSHA Challenge Program. While not officially a VPP, its intent is to assist workplaces that are interested in the VPP but need some help in meeting VPP requirements. The OSHA Challenge Program recognizes that there are many employers at different stages in the process of working toward implementing a successful safety and health management system and they require assistance in completing the process. The assistance they need is more than what is usually offered through the VPPPA Mentoring Program.

There are two tracks in the OSHA Challenge Program, one for general industry and one for construction. Participants follow a detailed three-stage roadmap that guides them to improve their safety and health management systems and work toward VPP status. The stages are used to mark the progress of the Challenge participant toward meeting the goals of the VPP in each of the elements of management leadership and employee involvement, worksite analysis, hazard prevention and control, and safety and health training. During each stage, the participant must document its achievements. The underlying progression expected is one from a reactive to a progressive method of managing safety and health. Progress may be demonstrated in several factors. Using the element of management leadership and employee involvement as an example, the expectations for each of the stages would be demonstrated by:

- *Stage 1:* Develop a safety and health mission statement and a safety and health policy statement with input from employees.
- *Stage 2:* Communicate the stage 1 statements and incorporate them into a new employee/contractor orientation.
- *Stage 3:* Take proactive steps to ensure the understanding of the statements by all employees and contract workers and that they become a routine part of regular communication.

During their participation in the OSHA Challenge Program, the participants receive assistance from other VPP sites or organizations that have volunteered to act as OSHA Challenge administrators or coordinators in developing their safety and health management systems. OSHA recognizes Challenge participants for each measured success and incremental improvement through the three stages of the program. OSHA provides incentives and recognition to Challenge participants at the completion of each stage to encourage their growth and implementation of a successful safety and health management system. Incentives may include access to compliance assistance and outreach, letters and certificates of recognition from OSHA, recognition on

OSHA's website, and priority scheduling for OSHA VPP onsite evaluations. The progress is reported to the Challenge Administrator who then provides the information to OSHA. As of November 30, 2008, the status of the OSHA Challenge Program was:

- 185 Participants
  - 96 in construction
  - 89 in general industry
- 22 completed stage 3
  - 18 in construction
  - 4 in general industry
- 10 achieved VPP recognition
  - 8 construction
  - 2 general industry
- 26 administrators
  - 14 for construction
  - 12 for general industry
- 114 coordinators
  - 75 for construction
  - 39 for general industry
- Total employees impacted
  - 28,338 in construction
  - 67,964 in general industry
- Site employees: 85,310
  - 25,797 in construction
  - 59,513 in general industry
- Contract employees: 10,992
  - 2541 in construction
  - 8451 in general industry

OSHA COOPERATIVE PROGRAMS

No discussion of the OSHA VPP would be complete without a discussion of the other OSHA cooperative programs. These include the Strategic Partnership and Alliance Program, the OSHA On-Site Consultation Program, and SHARP (Safety and Health Achievement Recognition Program).

Strategic Partnership Program

The Strategic Partnership Program started in November, 1998, as a means to enable OSHA to work with groups of employers, employees, employee representatives, and other stakeholders in order to encourage, assist, and recognize their efforts to eliminate serious hazards and achieve a high level of worker safety and health. Through this program OSHA works with the partners to recognize their efforts to eliminate serious hazards and achieve model workplace safety and health practices. Each partnership develops its own unique, formal agreement that establishes specific goals and strategies. The Partnership Program is available to work sites that fall under OSHA's jurisdiction. The Partnership Program process begins when an employer or other interested group informs OSHA that it is interested in working together with OSHA to ensure a safe workplace and working environment. A partnership agreement is written that will include the goals and the duration of the partnership. Partnership goals may include any one or more of the following examples: ultimate participation in the VPP, development of a comprehensive safety and health management system, development of ergonomic programs, development of industry-specific training programs, increasing focus on safety of non-English-speaking workers, increasing employee participation in the safety and health activities of the partner, and so on. Each of the goals must be measureable and be able to be validated for accuracy and effectiveness. Like the VPP, OSHA does perform a verification visit to the partner to confirm that hazards are controlled. Unlike the VPP, however, that verification visit is in the form of an OSHA enforcement inspection and citations and fines can be issued. Another difference from the VPP is the fact that OSHA will determine how many verification visits to perform each year.

General industry partnership participants may not receive any exemption from OSHA inspections, but they are eligible for special enforcement provisions so long as they are adhering to the partnership agreement. Those special enforcement provisions may include limited focus inspections.

Construction partnership participants may receive an exemption from programmed inspections after OSHA verifies the employer's safety and health performance through enforcement verification inspections. OSHA will determine how many enforcement inspections to perform each year based on the work activity of the participant. OSHA will perform at least one such inspection each year. During these enforcement verification inspections OSHA may issue citations and penalties for observed violations. In addition to assessing compliance with OSHA standards, the OSHA inspector will assess the participant's progress in meeting the requirements of the OSHA Strategic Partnership agreement and implementing an effective safety and health management system.

Alliance Program

Through the Alliance Program, OSHA works with groups committed to safety health, including businesses, trade or professional organizations, unions, and educational institutions, to leverage resources and expertise to develop compliance assistance tools and resources and share information with employers and employees to help prevent injuries, illnesses, and fatalities in the workplace.

Unlike the VPP and the Strategic Partnership Program, there is no evaluation or inspection component in the Alliance Program.

The Alliance participant must develop both short- and long-term goals for the program that are acceptable to OSHA. These goals must fall into one or more of three categories:

- *Training and Education* Examples include developing training and education programs and seminars aimed at reducing workplace hazards, providing the OSHA Training Institute with educational and training materials on specific safety issues upon request, and providing peer review of OSHA training curricula.
- *Outreach and Communication* Examples include sharing the most up-to-date ergonomic information for educational purposes, promoting participation in OSHA's cooperative programs, and providing information in Spanish and other languages.
- *Promoting the National Dialog on Workplace Safety and Health* By sharing data on safety and health hazards, participating in various forums and groups to discuss ways of improving workplace safety and health programs, and demonstrating the effectiveness of safety and health programs. OSHA representatives and Alliance Program participants have participated in numerous safety- and health-focused roundtables and other similar forums to discuss current issues in safety and health such as injuries and fatalities resulting from motor vehicle crashes and falls.

OSHA Onsite Consultation Service

The OSHA Onsite Consultation Service is a program through which OSHA funds state government agencies or state universities to offer free and confidential safety and health advice to primarily small and medium-sized businesses in all states across the country, with priority given to high-hazard work sites. The consultation services provided are separate from enforcement and do not result in penalties or citations so long as corrective actions are completed in a timely manner. Under this program, an employer may contact the state consultation service for an assistance visit.

The assistance visit may include a review of the safety and health programs in place, a noncompliance inspection of the workplace, industrial hygiene sampling, and an evaluation of the safety and health management system. The services also include employee training, programs development, hazard correction suggestions, and assistance in the development of comprehensive safety and health management systems. The OSHA Onsite Consultation Service activities are confidential and are not shared with OSHA, except in very rare circumstances.

In agreeing to the service, the employer also agrees to promptly correct all hazards identified during the onsite consultation visits. Failure to provide the state consultation service with any verification of hazard correction may result in a referral to OSHA based on noncompliance with the requirements of the consultation agreement.

Through the OSHA Onsite Consultation Program, the states have been empowered to work with companies to achieve recognition under the Safety and Health

Achievement Recognition Program (SHARP). SHARP provides recognition to small employers who operate an exemplary safety and health management system. SHARP preceded the VPP as the first OSHA recognition program.

Acceptance into SHARP by OSHA is an achievement of status that identifies the workplace as a model for work-site safety and health. Upon receiving SHARP recognition, the work site becomes exempt from programmed OSHA inspections during the period that SHARP certification is valid. The initial SHARP approval is 2 years with subsequent periods increased to 3 years.

To remain in SHARP, the employer must:

- Apply for renewal during the last quarter of the exemption period.
- Allow a full-service comprehensive visit to ensure that an exemplary safety and health management system has been effectively maintained or improved.
- Continue to meet all eligibility criteria and program requirements.
- Agree, if requesting a multiple-year renewal of 2 or 3 years, to conduct annual self-evaluations and to submit a written report to the state consultation program manager that is based on the elements of the 1989 Safety and Health Program Management Guidelines and includes OSHA's required injury and illness logs.

REFERENCES

1. National Safety Council (NSC), *Report on Injuries in America, Highlights from Injury Facts®*, 2007, Hasca, IL.
2. Public Law 91–596, Sec. (2) (b) (1).
3. Public Law 91–596, Sec. (2) (b) (13).
4. Margaret Richardson, "*Preparing for the Voluntary Protection Programs, Building Your Star Program*," 1999. Wiley-Interscience, Hoboken, NJ.
5. www.vpppa.org.
6. *Federal Register*, Notice 50 FR 43804.
7. *Federal Register*, Vol. 53, No.133, 7/12/88.
8. US Department of Labor Occupational Safety and Health Administration (OSHA), VPP Policies and Procedures Manual, CSP 03-01-003, Washington, DC, 2008.

2

THE BUSINESS CASE FOR VPP

INTRODUCTION

The Occupational Safety and Health Administration (OSHA) administers the nation's premier safety and health recognition program. Created in 1982, OSHA's Voluntary Protection Programs (VPP) recognize and partner with work sites that show excellence in safety and health. VPP sites are committed to effective employee protection beyond the minimum requirements of OSHA standards. There are three general requirements for participation in the VPP:

1. *An Effective, Ongoing Safety and Health Process* A comprehensive and effective safety and health process, involving all levels of employees, is the best way to prevent occupational injuries and illnesses. VPP work sites are expected to develop and implement a safety and health process that addresses the hazards present in the workplace and, in some cases, exceed the minimum requirements spelled out in the OSHA standards.
2. *Cooperation* VPP emphasizes cooperation and trust between the three key players in work-site safety and health: management, labor, and OSHA. The high level of cooperation found at VPP work sites complements OSHA's regulatory efforts.
3. *Good Performance* The safety and health management system is evaluated to ensure all applicable standards are addressed. Performance, in terms of

Preparing for OSHA's Voluntary Protection Programs. By Brian T. Bennett and Norman R. Deitch
Copyright © 2010 John Wiley & Sons, Inc.

occupational injury and illness rates, are also evaluated. VPP Star work sites must have injury rates that are below the national average for their particular industry classification. Merit work sites must be able to reduce their injury and illness rates below the national average for their particular industry classification within 3 years.

Good performance does not stop once the VPP onsite evaluation team has completed its review—continuous improvement is expected.

FOUR CORNERSTONES OF VPP

There are four cornerstones that the VPP has embraced as leading to a successful safety and health management system. The four cornerstones (elements) of the VPP safety and health management system are:

1. *Management Commitment and Employee Involvement* This cornerstone is believed by many to be the most important. First, management must be committed not only to the safety and health process but to the principles of the VPP. Management must ensure that the appropriate resources in terms of manpower, money, and equipment are allocated to provide the highest level of safety and health performance. Management must delegate responsibilities for safety and health to the appropriate personnel and hold them accountable for their performance. Management must also demonstrate their commitment by "walking the talk"; that is, actively participate in the safety and health process by attending safety meetings, presenting safety and health training, conducting inspections, and the like.

   The second part of this cornerstone involves the workforce. All levels of employees in every department within the facility must be able to actively participate in the safety and health process. There are a myriad of ways this can be accomplished, including:

   - Conducting inspections
   - Providing safety and health training
   - Participation on the emergency response team
   - Membership on the safety and health committee
   - Preparing or reviewing job hazard analyses
   - Performing incident investigations

   Almost any technique used to involve employees in the day-to-day operation of the safety and health process is acceptable, so long as their participation is active and meaningful. The intent of employee participation is to ensure that employees take ownership of the safety and health process.
2. *Worksite Analysis* This cornerstone involves the various systems and techniques used at the site to ascertain and quantify the hazards that might be

present in the work site and present a potential risk to the safety and health of employees, including:

- Job or process hazard analyses
- Industrial hygiene sampling
- Management of change procedures
- Hazard reporting and tracking system
- Incident investigation
- Routine inspections
- Occupational health care
- Review of the OSHA logs and leading indicators to identify trends

3. *Hazard Prevention and Control* Hazard prevention and control is a logical extension of the work-site analysis cornerstone. Once hazards have been identified, there needs to be a system to prevent them from occurring, or controlling the extent of the adverse effect if they were to occur. Hazard prevention and control typically includes programs such as:

   - Lockout/tagout
   - Confined space entry
   - Emergency response
   - Bloodborne pathogens
   - Hazard communication
   - Process safety management
   - Occupational health care

   The hierarchy of controls used to prevent and control hazards include engineering controls, administrative procedures, and personal protective equipment.

4. *Training* The final VPP cornerstone is training. This element includes all regulatory and job-specific training programs that have been implemented to ensure employees are fully trained and competent to conduct their tasks safely and efficiently. The training program should include all levels of employees in all departments. A comprehensive training program for contractors is also included in this cornerstone. VPP work sites often go beyond the minimum regulatory training required by the various standards and include other workplace and off-the-job training topics for employees as well as their families.

SELLING THE VPP TO YOUR ORGANIZATION

Companies are constantly searching for more ways to become more effective and efficient, to achieve the "upper hand" over their competitors. In today's global marketplace, competition between global and international competitors is intense.

In order to remain viable, companies must find creative ways to become more efficient and effective while cutting costs as low as possible. According to the

National Safety Council, accidents cost American businesses $170,000,000,000 per year. By adopting the principles of OSHA's Voluntary Protection Programs, a typical company will see a decrease of 53% in its accident rate. Although that is significant in the amount of not only financial cost but of pain and suffering avoided, there are even more benefits that can be realized by participating in the VPP. Many successful companies have taken the VPP model for safety and health and made that their business model. These companies have embraced the concept of going above and beyond the minimum requirements, of continuous improvement, and employee involvement to optimize all facets of their business.

As with any new initiative, thorough consideration must be given to the amount of resources, both in time and money, that must be committed in order to achieve the goal. Attaining VPP recognition will certainly require the expenditure of both types of resources. Management must thoroughly consider the business case, ensure they can realize a return on their investment, and commit the necessary resources.

Among the toughest concepts to "sell" to management to gain their support for the VPP include:

- Employee culture
  - Ensure everyone is on board with the importance of safety and health, as well as the importance of achieving VPP recognition.
- Going beyond compliance
  - OSHA standards are widely seen by safety and health professionals as the bare minimum requirements that must be met to ensure a safe and healthy workplace. The spirit of VPP is for sites to go beyond compliance; that is, beyond the minimum requirements.
- Inviting a regulatory agency into the facility
  - Why should we invite a regulatory agency into the facility when chances are unlikely we would be inspected in the foreseeable future? It may seem unadvisable to invite in an agency that can levy citations and fines, and perhaps lead to unfavorable public perceptions of your organization.
- Why continuous improvement?
  - Continuous improvement is not a new concept. For years companies have already adopted the concept of continuous improvement through their quality management systems. We know that excellent safety and health is a dynamic process, a goal that is always worked toward but never achieved. We must continue to explore new equipment and technologies and find new ways to encourage employee involvement, identify hazards in the workplace, implement effective controls to protect personnel, and devise new and interesting training programs.

Typically, it will take several years from the time the decision is made to pursue VPP recognition until the VPP Star work-site flag can be hoisted. The timing is directly proportional to the expenditure of resources: The more time and money that can be spent in the shortest possible period will expedite completion of the VPP requirements.

The key factors in how long it will take to achieve VPP recognition, and hence how many resources will be required, is based upon such issues as:

- The company's philosophy toward safety and health
  - Is safety and health an important core value, a priority, or neither?
- The safety and health and overall management culture of the organization
  - Does the organization embrace the concept of employee participation and empowerment?
  - Is it acceptable to have hourly employees lead improvement teams, chair safety and health committees, or present safety and health training sessions?
- The employee's participation in the safety and health process
  - Is there sufficient opportunity for active and meaningful employee involvement in the safety and health process?
- The sophistication of the safety and health process
  - Has the safety and health process been developed and implemented through the oversight of certified safety and health professionals?
  - Are all programs and procedures written and reviewed periodically to ensure they are up to date?
  - Have all work-site analysis and hazard prevention and control measures been addressed?
  - Are employee's current in all of their safety and health training?
- The physical condition of the facility
  - Has the physical plant been maintained to meet OSHA regulations and safety and health consensus standards as a minimum?

If you are truly committed to achieving safety and health excellence, the VPP will provide you with a clearer sense of direction and help you achieve your goals.

WHY PARTICIPATE IN THE VPP?

Many employers reason that barring a "catastrophe" (the hospitalization of three or more employees), a workplace fatality, or an employee complaint, the chance of OSHA conducting a programmed inspection at their facility is pretty slim, with the likelihood of an inspection occurring within the next 20 or 30 years unlikely. So why would an employer voluntarily invite OSHA into its workplace? Is it worth the risk?

A work site should not submit a VPP application until it is sure that it has a comprehensive, well-established safety and health process that not only complies with the OSHA standards but exceeds them in many areas.

A work site should apply to the VPP when it feels it has a mature safety and health process that involves all levels of employees, and is worthy of the recognition that VPP brings. However, work sites that are just beginning to develop their safety and health management systems can use the VPP application as a type of gap analysis that can serve as a template to develop an effective management system. They can then submit their application when they feel comfortable that they meet at least the Merit requirements. Achieving Merit status would then provide the additional motivation to complete their goals and ultimately achieve Star status.

BENEFITS OF THE VPP

There are many benefits that may accrue to an organization due to participation in the VPP, including:

- *Validation of Safety and Health Processes* A common concern among employees that have safety and health responsibilities at a work site is whether they are doing the right thing or doing enough to ensure not only that their employees are protected but that compliance with the OSHA standards has been achieved. By going through the VPP onsite evaluation process, the OSHA VPP team will evaluate each element of your safety and health process to ensure that it meets the minimum requirements of both the applicable OSHA standards and VPP.
- *No Penalty Inspection* One of the benefits of participation in the VPP is that if the OSHA VPP onsite evaluation team discovers violations of OSHA standards during the evaluation, no citations or fines will be issued. However, the expectation is that corrective actions will be implemented promptly.
- *Recognition* Achieving VPP Star status is the highest safety and health recognition that can be achieved. Star work sites have exemplary safety and health processes and are the best of the best. As an example, there are only approximately 2149 VPP work sites out of the approximately 7 million work sites eligible for participation. OSHA VPP recognition not only can raise employee morale but it can help foster acceptance in the community in which you operate as you have demonstrated your commitment to safety and health as a responsible member of the community.
- *Competitive Advantage* VPP work sites soon realize they have an edge against their competitors. Customers will benefit from lower costs due to the lower injury and accident rates. Quality will be higher as the same worker will be making the product every day since he or she is less likely to suffer an injury and be off from work. When measuring potential suppliers, the VPP may be the only criterion that differentiates your company from another. VPP status can make a company more marketable. Facilities that have been approved as a VPP work site can use the VPP logo and VPP designation on their products or marketing materials. However, caution must be used to avoid marketing a company or product as "OSHA Approved."
- *Higher Employee Morale* Employees that are fully engaged and participate in the safety and health process are more comfortable coming to work, knowing the chance of being injured or involved in an accident is extremely low compared to other work sites. In short, safe and healthy workers are happy workers!

- *Third-Party Verification* The OSHA VPP onsite evaluation team can serve as impartial, independent auditors, validating the effectiveness of your safety and health process. Unlike some other third-party program verifications, there are no application or certification fees associated with a VPP evaluation.
- *Removal from OSHA's Programmed Inspection List* VPP sites are removed from OSHA's programmed and targeted inspection lists. However, as mentioned previously, VPP work sites can be inspected upon OSHA's receipt of an employee complaint or the suffering of a catastrophic event. Violations observed during these inspections can result in the issuance of citations and fines, as appropriate.
- *Cooperative Environment between Labor, Management, and OSHA* Everyone wins in this scenario. Having a good preexisting labor–management relationship facilitates entry into the VPP. Achieving VPP status enhances labor–management relationships because of the openness, inclusiveness, ownership, and accountability that exist in VPP work sites. VPP work sites also enjoy a friendly, cooperative, and nonconfrontational relationship with OSHA.

BUSINESS CASE FOR VPP

Although there are many benefits for participation in the VPP, most of them are intangible and sometimes difficult to measure quantitatively. It is very difficult to sell ideas to upper management when they cannot reach out and touch the benefits. In one anecdotal VPP success story, to help convince upper management of the true value to the business of participation in the VPP, a study was conducted to relate VPP to the financial bottom line.

CASE STUDY OF VPP SUCCESS

Many VPP companies have realized various measures of success as a direct result of their commitment to OSHA's Voluntary Protection Programs. Following are just a few of those successes:

- After focusing safety and health activities on the VPP, the Department of the Navy realized a 4.8% increase in employee participation measured by the number of safety passports completed over an 18-month period. During the same period the rolling total case incident rate dropped about 33%.
- International Paper studied a mix of 174 similar VPP (50) and non-VPP (124) sites. It found that the VPP sites were significantly more successful in controlling hazards and reducing injuries and illnesses. VPP sites avoided about 18% of recordable events and 45% of lost time cases annually. If all of the sites were in the VPP, the company estimated that it could have realized a savings of $16,520,000 in workers' compensation costs.

- General Electric studied the effect of the VPP on its global safety and health management system effectiveness and uses the following reasons to support the use of the VPP for safety improvement:
  1. Structured process for improving safety performance
     - Verification of management systems that meet government expectations
     - Improved ability to find and fix hazards and compliance issues
  2. Requires buy-in from both management and labor
     - Team approach essential to improvement
     - Builds credibility with employees
  3. Proven results–real benefits
     - Decrease in injuries and lost time cases
     - Increases in productivity, quality, and employee morale
  4. Global standard—Mexico, Canada, and GE "Global Star"
     - Consistent process from Shanghai to Schenectady
  5. Builds positive working relationship and credibility with governments
     - Partnership with OSHA

In 1996, the author conducted a survey by soliciting interested VPP facilities. Fifteen VPP work sites across a variety of industries in the New York/New Jersey metro area agreed to voluntarily participate in the study. The purpose of the study was to determine if the VPP principles were extended to the business practices of the work sites, and if so did it lead to increased profitability. The results were, not surprisingly, extraordinary. Every measurable business indicator was significantly improved once the VPP principles had become ingrained in all aspects of the business.

Each of the work sites was asked to answer the questions based on its data for a period of 3 years before VPP participation against its data from 3 years after VPP participation:

1. *Workers' Compensation Insurance Costs* Costs were reduced approximately 55%.
2. *Injury/Illness Rates* Rates decreased by approximately 70%.
3. *Absenteeism* Employee absenteeism from work decreased from approximately 6 to 1.5%.
4. *Quality* Customer complaints due to poor product/package quality decreased approximately 85%.
5. *Hourly Participation in All Aspects of the Business* Approximately 90% of all hourly employees were actively involved in some business-related committee as opposed to less than 10% before the VPP process began.
6. *Union Grievances* Union grievances decreased an average of 75%.
7. *Budget Performance* One hundred percent of the work sites reported improvements in the actual costs compared to the budgeted costs.

8. *Profits* One hundred percent of the sites realized increased profits at their facility after VPP recognition.
9. *Accidents/Incidents* Work sites saw a decrease in accidents/incidents of 50%.
10. *Employee Perception* All work sites reported that the employee's perception of the company in terms of whether the facility was a good place to work had improved.

These results are attributable to the fact that the work sites extended the VPP principles of openness, management leadership, active and meaningful employee involvement, empowerment, and accountability to all employees across all facets of the operation. Each one of these indicators translates directly to increased profitability, which can easily be measured in terms of dollars and cents.

DOWNSIDE TO VPP PARTICIPATION

Participation in the VPP brings added value to an organization. However, it is not all peaches and cream. There are some recognized actual, potential, and perceived downsides to participating in the VPP. They may include some of the following issues:

- Many companies focus on the positive aspects of achieving zero lost workday injury rates (LWDI), whereas the VPP measures days away from work, restricted, and transferred (DART) incidents. That results in rates that are usually higher than the site's LWDI. The realization that the rate is no longer zero has been hard for some sites to accept.
- Work sites approved for Merit may realize a loss of commitment because of the perceived failure of achieving VPP Star recognition for the first time.
- Union work sites have encountered complications when the unions and/or management have placed the VPP on the negotiating table. That is significant since union support is a basic requirement for participation. VPP, and safety and health in general, is too important to all concerned to become a pawn in negotiations. Safety and health should not be subject to negotiation by either party.
- The VPP in itself does not add any significant costs to managing safety and health. However, most companies support the Voluntary Protection Program Participants Association by serving on various boards and committees and attending their conferences. They also support the principles of the VPP through the Special Government Employee (SGE) and Mentoring programs. Although those programs are voluntary, they are expected by OSHA and do add some additional costs.
- Companies may incur additional expenses to meet the additional VPP requirements of employee participation and to exceed the minimum OSHA standards. One example would be purchasing and installing automated electronic defibrillators although there is no OSHA standard that requires them.

SAFETY AS A CORE VALUE

Why are you reading this book? Most likely, because safety and health is important to you personally as well as your organization. Some organizations will tell you that safety and health is a key business priority, critical to their success. You will probably also hear that safety and health is important, and all safety and health rules will be strictly enforced. However, safety and health as a priority is just not good enough. And why is that? Quite simply, organizational priorities change based on what is important right now. Business priorities are important to our success—we need them. True, it may be that today safety and health is the number-one priority. However, due to business circumstances, tomorrow the number-one priority may be the budget or production or quality, and therefore safety and health moves down a notch or two in importance. So why is safety and health as a priority not good enough? Because priorities can and do change on a regular basis for a lot of different reasons, and in today's world they often change daily and in some cases even hourly! We cannot allow safety and health to be just a priority.

Safety and health is much too important to be a priority, something that is judged equal with all other business requirements. For those of us that are really serious about safety and health—those of us that want to be in OSHA's Voluntary Protection Programs—we must move to the next level and change our old philosophy that safety and health is a priority. To be successful, we must change both our organizational and personal culture to make safety and health a core value. Quite simply, that is really what the VPP is all about—always moving to that next level. Participation in the VPP helps organizations move to the next step and change the philosophy that safety and health is a priority. VPP sites believe that safety and health is a core value.

So what is a core value? Core values are those things that are so important to us, such as our family, that we would never compromise their significance by downgrading their importance to us. A core value mandates that we do the right thing, every time, even when it may not be comfortable for us. A truly effective safety and health process cannot and will not exist until everyone in the organization embraces safety and health as a core value. Once all employees, from the chief executive officer down to the newest employee, truly accept safety and health as a core value, will success will be in sight.

Now, we all know that talk is cheap—core values must be demonstrated continuously. What would happen if we were at work and saw an employee standing on a wheeled chair to reach something on a shelf? Most likely, something would be said to the effect of "stop that and don't do that again." Yet, how many of us have used a wheeled chair at home to reach something because no one is watching?

The bigger challenge is to make safety and health a core value, something that must be practiced 24 hours per day, 7 days per week, at home, at play, and at work, without exception. And we have an obligation to educate others, such as our co-workers and families, about core values—we must walk the talk about safety and health—and do what is right all the time, every time. We do not want people to do safety because someone is watching, we want them to live safety and make it an involuntary reflex without thinking about it.

SAFETY AND HEALTH AS A PROCESS

Another feature of VPP sites is the distinction between safety and health programs and safety and health processes.

As part of the concept of safety as a core value, VPP sites develop safety and health processes as opposed to safety and health programs. Programs are flavor of the month-type things—they are temporary things that have a defined life with a definite beginning and a definite end and typically address only short-term priorities and problems. As priorities shift, programs move up and down in importance. Conversely, processes are core values that never lose their importance and stature to the organization or its employees. Processes are systems that go on forever, are a reflection of the organization's core values, are ingrained in the organization's and individual's everyday activities, and are continuously improved.

THE NEXT STEP—VPP

How would your management react if the accountant at your company announced at the next staff meeting that he was inviting the Internal Revenue Service—the IRS—in to the office to review all of the company's financial records for the past 5 years? Most likely, that accountant would be enjoying a well-deserved vacation in a very quiet location for a long time.

It may seem counterintuitive to invite a regulatory agency into your facility to scrutinize your facility and records, inspect all aspects of your workplace, and have unimpeded access to your employees to see and hear all the good (and not so good) things you may be doing. However, this concept is not so far out once your organization has embraced safety and health as a core value, and its precept of continuous improvement.

Can you have a good and effective safety and health system without being a VPP site? Absolutely. But, if you want to have an even better safety and health system, go beyond compliance and provide even more protection to your employees, if you buy into this concept of the importance of core values versus priorities, and you want to integrate core values into your entire business process and move to that next step, then the VPP is what you need.

WHAT IS THE VPP?

So what is the VPP? OSHA will tell you it is the Voluntary Protection Programs. The VPP is a recognition program for those employers who have exemplary safety and health programs. OSHA VPP sites typically have accident rates that are 53% below the national average for their industry. VPP work sites are removed from OSHA's programmed inspection list. There are approximately 7 million work sites in the United States that are eligible to participate in the VPP, yet only 2149 have met the stringent VPP requirements.

And yes, all of that is good and all of that is important, but it only scratches the surface. Even more important is that VPP companies have taken that next step and acknowledged that safety and health is a core value in their organization. VPP companies are true safety zealots, those that are at the front of the pack, true leaders at the cutting edge of safety and health.

I would go so far as to say these companies would tell you they have learned that VPP stands for something else, which is much more important to them in regards to their safety and health systems: Value, Performance, and Passion.

VALUE

For those companies that have embraced safety and health as a core value, inviting OSHA in to their facilities does not seem like a bad idea. Remember the VPP does not stand for the Voluntary *Perfection* Programs.

Since we have already established that safety and health is a core value, then it is only natural that we want to make our safety and health process stronger and even more ingrained in our organization. And that is exactly what the VPP is—not just an award—but a continuous process reaching toward excellence. So what can the VPP really do for us, what is the real value of participating? For those of you that are not yet a VPP work site, let me tell you about the top 10 values a typical VPP work site will enjoy:

1. Decreased insurance costs—If you have a decrease in injuries and accidents, you will see a corresponding decrease in workers' compensation, property, and liability insurance costs.
2. Improvement in quality—If we are successful in extending the VPP concepts to other areas of the business and having employees embrace the concept that quality is a core value and devote the same level of attention, just like safety and health, you will see a marked improvement in the overall quality of your employees work.
3. Decrease in absenteeism—When employees feel safe, when employees feel involved and empowered, when employees have a say in the business, they feel good about coming to work and look forward to contributing to the overall success of the business.
4. Improved employee perceptions—Employees morale and attitudes toward the company are enhanced when there is a safe and healthy workplace. Happy employees equates to more efficient and productive employees.
5. Better community relations—Some work sites have a history of poor performance that directly and adversely affects the neighborhood in which they operate—whether it involves incidents such as fires or hazardous materials releases, injuries to family members or friends, pollution, and the like. Once VPP caliber systems are in place, these problems go away and the company enjoys a new and improved relationship with the community. The facility is now seen as a responsible member of the community, one that

respects and has taken the steps necessary to protect the people and environment that surround the facility as well as the people that work at the facility.

6. Validation of the safety systems—Most safety managers share common concerns: Am I in compliance with the OSHA standards; have I done the right things to protect employees; is there anything I forgot that can lead to an injury. The VPP process involves a thorough and independent unbiased third-party evaluation of the safety and health management system. Surviving the scrutiny of a VPP on site evaluation is a validation that the safety systems are indeed exemplary (but not perfect).
7. Cooperative relationship with OSHA—In the VPP realm, OSHA is now seen as a partner, a resource, a consultant to facilitate positive change rather than the bad guy showing up with a ticket book and issuing citations and fines. The VPP site can rely on OSHA to assist in resolving safety and health issues without fear of penalty, and OSHA has additional expertise that can be tapped to assist it in furthering its mission through the VPP mentoring and SGE programs.
8. Improved employee relations—The VPP requires an excellent labor–management relationship, one that facilitates open, two-way discussions to resolve issues that relate to all aspects of the business. As part of the VPP process, trust between labor and management has already been built. For unionized sites, this translates to a significant decrease in the number of grievances, as issues are discussed in a nonconfrontational manner to the mutual satisfaction of all parties.
9. Increased profits—When all of the VPP elements are fully implemented, a company's profits only have one place to go and that is up ! Therefore, effective safety and health management systems directly benefit all stakeholders, a true win–win situation!
10. Lower injury and illness rates—Nobody wants to get hurt or sick, and no boss wants to see their employees injured or killed. Because of the sophistication of their safety and health management system, VPP members typically enjoy injury rates that are an average of 53% below the industry average for their particular industry.

In today's global marketplace, with competition being extremely tough, companies must find and implement each and every opportunity available to streamline their business and make it more effective, efficient, and profitable. Don't you think the Voluntary Protection Programs offer significant value to help improve your bottom line?

PERFORMANCE

Every day workplace injuries and accidents in the United States leave a trail of death and destruction to the tune of $170 billion annually. The VPP concepts help companies eliminate injuries and accidents and make businesses more competitive and more profitable. However, achieving VPP recognition is not easy, and it is not meant to be. VPP recognition must be earned. In order to bring value to an organization, the VPP process must maintain its standards and integrity, without compromise. VPP status is not bought, it is not given to those that are our friends, and it is not given to those that are close to being excellent. VPP recognition is given only to those work sites that fully meet all requirements for Star, Star Demonstration, or Merit recognition.

Good performance is important to us. We learned that as children when we received our report card from school. We can grade our safety and health process using the same system. If we were just making it in school, doing the minimum, we probably received a grade of D—passing but just barely. That would be the grade a facility that was just complying with the OSHA standards would receive. If a facility was not in compliance with the OSHA standards, it would clearly receive a failing grade, or an F. For those sites that have started on the path to safety and health excellence and are a SHARP (Safety and Health Achievement Recognition Program) site, they would receive a grade of C—average or slightly above. VPP Merit sites—those facilities that are starting to set themselves apart from everyone else—would receive a grade of B. And VPP Star sites, the best of the best, would receive a grade of A.

Speak with those sites that have gone through the VPP process. It is an arduous task that often takes years to achieve. The safety and health culture may need to change to embrace the concept of core values; safety and health systems may have to be developed, implemented, and enhanced; engineering controls and hazard control systems may need to be implemented to enhance worker safety and health; and finally safety and training programs may have to be developed and presented to ensure all employees are aware of the workplace hazards and corrective actions necessary to avoid injury. The VPP should be seen as a challenge to do all of the right things in regard to safety and health and cause no harm to our employees, the environment, or the neighborhood in which we operate.

The key to success in the VPP is performance. VPP cannot be achieved without outstanding performance by all employees. There are minimum, objective standards that must be met, such as having a 3-year average total case incident rate and days away, restricted and transferred rate below the national average for your industry to qualify as a VPP Star work site. But the real challenge in achieving exceptional performance begins and ends with management leadership and employee involvement.

Management must be committed to the concept of safety and health as a core value, and must set the tone for the business and all employees to meet. Successful employee involvement often involves empowerment of all employees to be actively and meaningfully involved in all aspects of the safety and health process. Hourly employees, those on the front lines working on the floor, are often found leading safety and health committees, working on safety improvement teams, and presenting safety and health training to their co-workers at VPP work sites.

Leading, cutting edge companies often take the VPP model and its precepts and expand them into all aspects of the business. These companies have discovered the VPP framework, and the success that it leads to in regards to safety and health, can have the same results in terms of budget performance, quality, productivity, initiative, and development of new techniques to enhance the business. Quite simply, VPP companies are models of excellence, not only in safety and health but usually in other business areas as well.

Is there a cost to achieve VPP? Absolutely. However, according to a study done by the Voluntary Protection Program Participants Association, it is estimated that for every dollar invested in safety and health, there will be a return of $4–6 dollars. That sounds like good math to me! Most companies are happy when they receive a 15–25% return on investment.

PASSION

Safety and health is often compared to a religion. You must have passion and faith in order to achieve rewards.

Employees at VPP work sites are passionate about safety and health and are passionate about the VPP. They enjoy the prestige that comes with being "best in class." They have the passion to succeed. They have the passion to do the right thing. They recognize that the VPP is the only *meaningful* global recognition program for exemplary safety and health performance. VPP will bring bona fide peer recognition to your facility and help build the passion for safety and health in your employees.

Passion means safety and health is ingrained in everything we do; it has now become second nature, an involuntary reflex. Why are we passionate? What led to this conversion in just a few short years? We have often lived through the tough times when injury and accident rates were high. We saw friends being injured and in some cases killed. We saw jobs jeopardized by poor safety and health performance that had adverse effects on other aspects of the business. We saw increased costs, declining productivity and quality, poor employee morale, reductions in force, companies going out of business, or facilities being closed due to spiraling costs. The savior was the Voluntary Protection Programs. We embraced the VPP philosophy of safety and health as a core value, of management leadership, and employee involvement. We saw the light at the end of the tunnel and it was VPP.

Once the VPP concepts were embraced and implemented, things turned around rather quickly. Employees are in a much better world now that VPP recognition has been achieved. Injuries are down, and productivity, quality, morale, and profits are up. Employees are excited to come to work and make a difference every day because they are involved and empowered in all aspects of the business. There is now pride in the workplace—pride in being able to fly the Star or Merit flag; pride in telling their family and friends they work in a VPP work site; pride to be at the top of the class; pride in telling their friends who have not yet been converted to the VPP way that they are a VPP employee.

Everyone has passion about safety and health for different reasons. For the authors, our passion comes from involvement in the community as volunteer emergency responders. Every day, 16 employees are killed in workplace accidents in the United States.

We have seen first hand the devastation a workplace incident can cause and what happens when safety and health is a priority. We have pulled lifeless bodies from confined spaces because companies did not have air monitoring, an entry permit, or any of the other precautions because time was the priority and the job had to get done. We have treated workers horribly burned in workplace fires due to inadequate hot work procedures because money was the priority, and there wasn't someone available to write a procedure. We have tried to resuscitate heart attack victims who could have been saved had fellow employees been trained in CPR or the company invested in an automated external defibrillator (AED) because time was the priority, and they were too busy to do the training. We have spent hours extricating employees that were swallowed up by machines that were not guarded or not locked out because production was the priority. We have had to pick up pieces of employees blown apart in an explosion because a company did not feel it had the resources to implement a process safety management program because product quality was the priority. We have had the unpleasant task of telling an employee's survivors that their dad, son, or brother will not be coming home tonight because something else was always a higher priority than safety and health.

But it doesn't stop at work—off-the-job fatalities far outweigh occupational fatalities. It could be a motor vehicle accident, a fire in the home, a child falling into an unprotected swimming pool, or someone falling off a chair used as a ladder and striking his or her head—the result is the same, the consequences are the same, and the survivor's pain is the same. Years of experiencing this has led to our passion for safety and health, why we have embraced safety and health as a core value, and why we believe VPP is the mechanism to achieve excellence in safety and health, both at work and at home. Why are you passionate about safety and health?

DO THE RIGHT THING

The VPP is an excellent way to elevate your safety and healthy process to the next level. The VPP provides many tangible and intangible benefits to all stakeholders—management, labor, OSHA, and the community. Companies that have participated in the VPP have seen enhanced performance not only in safety and health but other business metrics as well.

The VPP brings value to each of a company's stakeholders:

- *Employees* Have a clearer sense of direction toward safety and health, improved communications, and an overall safer work site.
- *Owners* Will ultimately see lower costs and higher profits.
- *Customers* Will have increased satisfaction, fewer problems with reliability and delivery, and higher quality, better products and services.
- *Community* Will benefit from a safer, more stable employer and a responsible community member.[1]

The question now is, are you passionate about safety and health? Do you want to share in all the benefits of VPP? Are you committed to embracing safety and health as a core value? Are you willing to begin the journey down the road to becoming a VPP work site?

REFERENCE

1. B. Bennett and N. Deitch. OSHA's VPP. *Professional Safety*, **52**(12), 24–31 (2007).

3

PREPARING AN OSHA VPP APPLICATION

Before the decision can be made to participate in the VPP, labor and management must agree that it is the right thing to do for the business, management, and the employees. The decision by management to participate in the VPP should be based on consideration of all factors discussed earlier. Specific attention should be given to the expected benefits and the potential additional costs to manage and sustain the program. For the purposes of this discussion, we will use the term *employees* to represent the hourly, nonexempt workers at a workplace. *Managers*, *directors*, and *supervisors* will collectively be considered management.

All employees and management must be informed about the details of the VPP so that they may be able to make an informed decision to support or not support VPP participation. Whereas nonorganized workforces can demonstrate their agreement by general consensus, union work sites must obtain a formal notification from the local union representative that the union either supports the application or at least does not object to it.

THE VPP APPLICATION PROCESS

The general flow of the VPP application process is as follows:

- Educate your employees about the VPP.
  - You should present the VPP concepts, benefits, and downfalls to your management team and employees to gain their support for the program.

Preparing for OSHA's Voluntary Protection Programs. By Brian T. Bennett and Norman R. Deitch
Copyright © 2010 John Wiley & Sons, Inc.

Without the full, unconditional support of all involved parties, the VPP process cannot proceed.

- Select a team to write the VPP application.
  - An individual or team should be selected to prepare the VPP application.
- Revise the VPP application.
  - Once the team or individual has completed writing the draft VPP application, it should be reviewed to ensure it is complete and thorough.
- Submit the VPP application.
  - Once the VPP application has been reviewed and finalized, it should be submitted to the federal OSHA VPP Manager or the state plan VPP Coordinator.
  - The VPP Manager or VPP Coordinator will inform you within 15 business days that your VPP application has been received.
  - The VPP Manager or VPP Coordinator generally replies back to you within 30 days letting you know whether the application is complete and has been accepted.
- Provide additional information to the VPP Manager of VPP Coordinator.
  - If the VPP application is incomplete or requires clarification, the VPP Manager or VPP Coordinator may request you supply additional information. You must reply back with the requested information within 90 calendar days.
- Schedule the onsite VPP evaluation.
  - Once the VPP Manager or VPP Coordinator has formally accepted your VPP application, he or she will contact you to schedule the onsite VPP evaluation. The evaluation must be conducted within 6 months of approval of the application.

VPP EDUCATION

A basic requirement of the VPP is that employees have a basic understanding of the VPP—what is in it for them, and what is in it for the company. Education about the VPP can be presented in various ways. There is some informational literature on the OSHA website, and there are numerous vendors that market VPP-related literature and marketing services. The various VPPPA conferences also offer information about the program.

The more effective programs involve many employees to develop and present promotional activities. These may start with an orientation for all managers and employees and be reinforced with routine formal and informal training programs. One successful program at a large facility with over 3000 employees started with a video that provided general information about the VPP and was required viewing by all workers at the facility. That was then followed up with small staff meetings and other meeting opportunities, such as bag lunches and formal detailed VPP training. The facility VPP leaders created a theme for the program and that became the common thread for all VPP activities. Posters, trinkets, gifts, games, and safety days were all based on the theme and all employees were exposed to the information.

Consultants that specialize in the VPP can also provide valuable assistance throughout the entire process.

THE STARTING POINT: COMPLETE A GAP ANALYSIS

Once a decision to move forward with the VPP process has been made, the facility should undergo a comprehensive gap analysis of the entire safety and health management system. All elements and factors considered by the VPP must be reviewed for their effectiveness. This may be performed by the safety and health staff or a multidisciplinary gap analysis team (GAT). It may also be performed by corporate teams or consultants. The team approach is generally preferred because it would more likely result in a more objective evaluation. The team can be comprised of employees from all levels and departments, as well as outside resources such as corporate staff, mentors, or consultants. Selection of GAT members should be volunteer based, but the final selection of the team members should include a review of several pertinent criteria including technical expertise and knowledge of the safety and health systems.

The primary task of this team is the completion of the gap analysis. The gap analysis is an audit of the site's existing policies, programs, and processes measured against the required VPP elements and subelements. Upon completion of the gap analysis, a corrective action plan that lists all of the shortcomings should be created. The corrective action plan should assign specific action items to identified personnel along with a targeted completion date.

The gap analysis team should meet on a regular basis to track the status of the corrective actions, ensure progress is being made, and lend their support when necessary.

The gap analysis should be started as soon as possible, to identify shortcomings in the safety and health management system. This is critical since a requirement in the VPP is that all required elements and subelements be in place and functional for a minimum period of one year before Star status can be awarded.

The authors support the direct involvement of employees in the gap analysis process and they should be an integral part of the GAT. However, the completion of a gap analysis is extremely technical and must address not only how the various safety and health activities are performed, but it also must review the facilities compliance with all applicable OSHA rules, regulations, and expectations. Therefore, the team must include safety and health specialists and have availability to other technical resources. Several successful gap analyses have used this team approach and the subsequent OSHA VPP evaluations validated the accuracy of the teams' work.

Those facilities with a formal safety committee may consider using that committee to perform the gap analysis. We do not recommend that for several reasons. First, the VPP would most likely not be in the charter of the safety committee. Second, adding the additional responsibilities for the gap analysis would more likely than not overtax the typical limited resources of the committee. Third, using the safety committee precludes the involvement of additional employees in the safety and health process. Having said that, the GAT may be considered as a subcommittee of the safety committee.

Typically, gap analysis recommendations fall into one of several categories:

- *Minimum requirements*: These recommendations should be completed immediately as they are the minimum requirements that need to be met for participation in the VPP.
- *Compliance issues*: These recommendations involve corrections that must be made in order to achieve compliance with OSHA standards.
- *Enhancements*: These recommendations involve things that can be done to enhance the safety and health systems and elevate it to VPP Star-level quality.

It is acceptable to submit a VPP application (and perhaps even be recognized as a VPP work site) with open action items. Consideration would be given to the number and type of open action items, along with their severity. For action items that require long lead times for funding or equipment design and installation, an action plan may be sufficient to allow the VPP process to continue.

WRITING THE VPP APPLICATION

Once you have decided that the VPP is a good match for your company and employees and you have addressed the corrective actions identified in the gap analysis, you are ready for the next step in the VPP process: the application has to be written and submitted. Although the application is fairly straightforward, there are many different approaches that can be taken to actually write it. This section will discuss the application and suggest some of those ways that other companies have found successful.

The first thing that must be understood is what the application is and is not. The VPP application can be very simply described as a comprehensive narrative summary of how you manage each of the four major elements of your safety and health management system and each of its subelements. Using the OSHA Guidelines for Safety and Health Management Systems, the VPP Policies and Procedures Manual, and the *Federal Register*, the four elements are:

- Management Leadership and Employee Involvement
- Work-site Analysis
- Hazard Prevention and Control
- Safety and Health Training

The number of subelements is variable based on each company's specific operations, processes, organization, chemicals used, and size. Although you may have heard about the "19 VPP Elements" there are actually 28 elements and subelements. All of the elements and subelements are defined on the OSHA application. See the VPP elements in Exhibit 3.1.

Writing the VPP application can be a daunting task. To begin with, as part of the Paper Work Reduction Act, OSHA has estimated that the process of writing the VPP application takes about 200 hours. For any safety and health manager, that is a lot of time that would take away from his or her very full agenda. It may also have a detrimental effect on the overall efficiency of the safety and health management system in the workplace. That would tend to result in a somewhat negative attitude toward the entire VPP process.

In fact, the writing of the application should not be that intimidating. There are many tools and resources available to help facilitate the process. The first place to start is to create a dialog with the regional OSHA or state plan VPP Coordinator. He or she can be identified through the VPP link on the OSHA website (www.osha.gov). In addition to informing the coordinator of your decision to submit an application, he or she should be able to provide you with an application template and provide some assistance. The coordinator could also refer you to other sources of information about the application process. Most importantly, he or she will let you know what is expected in a VPP application. Some VPP Coordinators prefer a substantial amount of detail and attachments in the application, and some prefer less detail and limited attachments. The general approach has been that the application should "bedazzle the reader with brilliance and brevity; not burden the reader with bulk."

Once you have received and reviewed the application format, the work-site application lead person must develop a method and plan to get the application written. There are several methods that should be considered. We will review each method and discuss the benefits and issues of each.

The first and most obvious method is for the VPP application lead person to write the entire application without any, or just limited, assistance. The benefit of this method is that the application lead person is usually the site safety manager or similar position that is intimately familiar with the entire system. Offsetting that benefit is the time that would be required to write the application. Most safety managers can ill afford to devote about 200 hours to the application with all of their other responsibilities. Writing the application would thus result in a detriment to the ongoing safety and health activities at the workplace as the focus would be on writing the VPP application rather than managing the day-to-day activities. The challenge to balance the writing of the application with the time required in managing the safety and health management system would also most likely result in a delay in the completion of the application. Another potential deficiency with this method is that as the primary owner of the safety and health management system (System) the safety manager may not be as objective in the description of the System as necessary. This is important because the process of writing the application provides another opportunity to critically evaluate the effectiveness of the System. In fact, many that have begun to write a VPP application have recognized that they have deficiencies in the System and have delayed submitting their applications until those deficiencies are corrected.

Most safety managers that have written their own VPP applications have recognized that they could have made more productive use of their time. They have informed the authors that they would not do it again and would not recommend that method to others. We believe that this method may be most appropriate for those smaller work sites that have limited access to other internal or external resources to assist them. Fortunately, their Systems are usually not as complex and detailed, thus reducing the burden somewhat.

Another method to complete a VPP application is to assign it to a team. This method also has both pros and cons associated with it. The main benefit is that it spreads the workload among several staff members and employees. It also provides opportunities for more employees to be actively involved in the safety and health management system and the VPP process at the workplace. The first issue with this method is the selection of the VPP application team. It may be difficult to get volunteers to join the VPP application team and directing them to participate may not result in the positive benefits expected. Another concern is that those who volunteer may not have the requisite knowledge of the System and the application VPP and/or the technical ability to write their assigned section of the application. Therefore, the selection and assignment of the VPP application team must be given very careful and thoughtful consideration. The ideal VPP application team should consist of a mix of both hourly and salaried employees and should represent different areas of the work site. That would more likely result in a balanced approach to the final product.

The VPP application team would have to receive detailed information about the VPP program and process and agree that it is the right program for the work site to demonstrate its commitment to the safety and health of the workers. There are several sources of information that are available. The first place to go is the VPP link on the OSHA website. We also suggest contacting other local VPP sites that may assist as a mentor. Information about mentors is available from the Voluntary Protection Programs Participants Association (VPPPA) (www.vpppa.org) or the regional chapter, which can be reached through the same website.

Mentors are representatives of current VPP sites that have been through the VPP process and are familiar with the challenges of the application process. They have volunteered to assist other companies to work through the process to become a VPP work site. The amount of assistance that they provide is not proscribed in any manner and may be worked out with the mentor directly.

To ensure success using the team approach, management would have to allow sufficient time for the team to meet to discuss their assignments, to collect the necessary data and information, to interview the principals involved in the specific function, to actually write their sections, and to meet regularly to discuss their progress and the overall status of the process. They will also need to be provided with the necessary computer equipment to write the application. Fortunately, most VPP application templates lend themselves very well to the principle of cut and paste. This method facilitates the building of a complete application from the individual parts. Also, fortunately, OSHA is not interested in the writing style in the application—only the content. A third approach to writing the VPP application is using a VPP mentor through the VPPPA. This approach is not widely available since many individuals and companies that have agreed to mentor other work sites on their VPP application process limit their involvement to providing assistance and suggestions regarding the publicity of the VPP among the workers. Some mentors may go so far as to assist the mentee in the review of the various safety and health control programs, but even that is relatively rare. Companies are becoming more reluctant to become involved in the review of such programs because of the potential liability should they overlook something in their review.

Mentors also have not been involved in the actual writing of a mentee's VPP application. Since the mentor is probably not that familiar with the applicant's entire safety and health management system, organization, and other pertinent details of the mentee's work site, mentors would not be in a position to answer the questions contained in most application templates. Mentees would also possibly be reluctant to provide details of their manufacturing process to ensure the security of their proprietary information that cannot be shared with other companies.

A final method of completing the application is to hire a consultant familiar with the VPP to either assist the company authors (individual or team) with the application process or to write the application themselves. Whereas this would result in a direct cost to the company, it would result in an acceptable VPP application and save the indirect costs associated with writing the application in-house.

Consultants may have the benefit of being very familiar with the VPP and the application process. Experienced consultants can often produce an acceptable VPP application in significantly less than 200 hours. When combined with a gap analysis of the work site's safety and health management system and an audit of the workplace, the consultant can offer the advantage of providing a more objective view of the company's System. They would be in a very good position to identify areas of improvement and to make suggestions for improvement. They could also suggest a reasonable timeline for the VPP process. Caution must be exercised to ensure that the consultant is well versed in not only the gap analysis process but the requirements of the VPP application and the expectations of OSHA regarding the application, as discussed earlier.

The authors suggest the use of a team approach for the development of the application. This approach enables more employees to be involved and reduces the burden on the primary person responsible for the application. Many successful applications were written using this approach. Most of those work sites created a committee, usually referred to as the VPP steering team or committee. They were charged with developing not only the necessary VPP awareness and publicity campaigns but also with developing a complete application. This is not intended to imply that the team/committee actually wrote the application. They may have written some sections of the application of which they were knowledgeable. One example of this would be an hourly employee discussing how hourly employees are involved in the safety and health activities at the work site. Another example would be for the emergency response team to discuss the emergency response section of the application. Subject matter experts may be used to write specific sections of the application, such as a chemical process safety management (PSM) leader could write the PSM addendum to the application. Other sections and suggested assignments include:

- Contractor activities assigned to the procurement group
- Personnel responsibilities and accountability and discipline to the human resources group
- Preventive maintenance to the maintenance group
- Hazard analysis to the safety group

- Occupational health to the medical group
- Training to the training group

The general first section of the VPP application is usually best written by the safety lead of the work site. The first section deals with the company description, contractor activities, injury and illness records, and the North American Industrial Classification System and Standard Industrial Classification codes.

It would be the VPP application team leader's responsibility to ensure that all those with specific assignments to write sections of the application complete them within the prescribed time frames. The team leader could then assemble each section or may assign someone else to build the final application. Once assembled, the application would have to be reviewed by the entire team and edited for completeness and accuracy. An inexperienced individual or team can expect to spend 3–6 months to develop and review a VPP application.

Once the application has been written and reviewed, the next task is to assemble the required and suggested attachments. This is where the initial contact with the regional OSHA VPP coordinator will demonstrate its importance. The OSHA VPP Coordinator will inform the VPP application team what attachments are required. This will vary from region to region and possibly from one office to another within the same region. As discussed earlier, some OSHA VPP Coordinators require only those attachments that are specifically required by the VPP Policies and Procedures Manual and the VPP *Federal Register*. Other VPP Coordinators expect numerous attachments as evidence of the activities described in the application.

APPLICATION ATTACHMENTS

The VPP requires the following attachments to be included with the application:

- Statement of assurances
- Union statement of concurrence
- Organization chart illustrating the line of responsibility of the lead safety and health position
- Details of the respiratory protection program or the entire program
- The most recent annual evaluation
- The table of injuries and illnesses for the last three full calendar years, with information about the current year to date

Additional attachments may include:

- Actual examples or templates for annual performance evaluations for salaried and hourly employees
- Examples of disciplinary actions
- Examples of inspection reports
- Training matrix

- Safety committee meeting minutes
- Examples of accident investigations
- Examples of recognition programs and awards

Some OSHA VPP Coordinators may also request that all hazard control programs be attached. However, that is the exception rather than the rule since those programs will be reviewed during the onsite evaluation. It is also recognized that those programs submitted with the application may be revised by the time of the evaluation, making them obsolete.

REVIEW AND SUBMITTAL OF THE APPLICATION

When the entire application is completed and reviewed, a VPP mentor may be used to review it to determine if it would meet the requirements of OSHA. Caution must be used when asking a mentor to review the application. During its evolution, the VPP has seen several revisions to the application process. If the mentor was not a recent applicant, he or she may not be aware of the current requirements. Furthermore, even recent applicants may not have submitted applications that met the expectations of the VPP Coordinator. This may result because many in OSHA will accept applications that do not follow the recommended format as long as the application contains all of the required information.

Many applicants have delayed the submission of their applications because they were not sure that the applications were complete or detailed enough. Others have asked OSHA if they could submit the application in draft form for a review before a formal submission is made. Both of these actions usually result in unnecessary delays. One of the most frequent causes for delay in the submission of the application is second guessing its contents. In the experience of the authors, OSHA's review of most applications result in a request for additional information or clarification. There is no prejudice applied to initially deficient applications. The applicant is given 90 days to respond to requests from OSHA for such additional detail or clarifications. Only if the applicant does not provide the requested information will OSHA return the application after 90 days. Even if that should occur, the entire application may be resubmitted without prejudice at any time.

The authors do not recommend the submission of a draft application. Since OSHA reviews all applications in detail, all applications may be considered as a draft until they are formally accepted by OSHA. Labeling the application submission as a draft may imply an uncertainty as to its contents or completeness or the readiness of the work site to participate in the VPP.

RETURN OF THE APPLICATION

It is possible that the regional OSHA VPP Manager or VPP Coordinator may contact you for additional information or clarification. This is somewhat routine and is not indicative of a problem. OSHA allows the applicant up to 90 days to submit the requested material and information. The request for additional material or information may include information such as:

- Clarification of how the applicant ensures that the entire workplace is inspected quarterly
- Specific details of the safety and health responsibilities of managers and supervisors
- Details about how managers and supervisors are held accountable for the performance of their safety and health responsibilities
- Examples of completed job safety analyses
- A copy of the annual sampling plan
- Examples of accident investigation reports
- Inspection checklists

The requested information and materials should be submitted promptly to move the application approval process forward. The VPP Policies and Procedures Manual allows the applicant 90 days to submit the requested material. If the material is not received by OSHA within 90 days, the VPP Manager must return the application as incomplete. The applicant may be resubmitted at any time after the requested material is added.

If OSHA determines that the applicant will clearly not qualify for the VPP, it will ask the applicant to withdraw the application within 30 days. If the application is not withdrawn, the VPP Manager must return the application with a letter indicating the reasons the application was denied by OSHA. The VPP Manager must also forward a copy of that return letter to the OSHA national office. Some reasons an applicant will clearly not qualify for the VPP include:

- An upheld willful OSHA citation within the previous 3 years.
- Failure to obtain union agreement with the application.
- Injury/illness recordable rates that are significantly higher than the industry average rates.
- An open enforcement investigation.
- Pending or open contested citations or notices under appeal at the time of application.
- Affirmed willful or 11(c) violations during the 36 months prior to application. An 11(c) violation refers to section 11(c) of the OSH Act that protects employees from any act of discrimination as a result of filing a safety and or health complaint with the employer or OSHA.
- Unresolved, outstanding enforcement actions such as long-term abatement agreements or contests.
- OSHA history pertaining to a non-VPP work site of the same company will not adversely affect VPP participation, unless it is determined that a corporate

decision, program, or policy that applies to all company work sites does not meet OSHA standards.
- Significant negative interactions with OSHA during the previous 3 years.

The applicant can also voluntarily request that OSHA return the application without prejudice. This may be a result of a change of management or financial situation of the company. It may also be a result of a reassessment of the readiness of the applicant to undergo an OSHA VPP onsite evaluation. OSHA will honor such requests within 10 working days of receipt. Voluntary requests for the withdrawal of an application are considered without prejudice, and the application can be resubmitted at any time.

THE APPLICATION BY SECTION

The VPP application will now be discussed section by section. We will discuss each section to describe what OSHA is looking for and how to answer the questions. The section will also provide instruction on how to complete the required tables to report the work site's and applicable contractors' injury and illness rates.

A. General Information

This section contains information about the work site, the facility manager, the VPP contact person (the one with the responsibility for the application process), any unions represented, the products produced, and the industry classification. The detailed information requested in this section is entered in a narrative format. A copy of a template for this information follows (with the italics used to define the information requested by OSHA as part of the VPP application, and the authors' explanatory notes follow in regular type). The referenced tables of injury and illness rates for company and contractor employees are discussed in greater detail at the end of this section.

1. Applicant

Site Name:

Site Address:

Site Manager:

Title:

Site VPP Contact for OSHA Correspondence:

Title:

Phone Number of VPP Contact:

E-mail Address of VPP Contact:

2. Company/Corporate Name

Name (if different from above):

Address:

VPP Contact (if applicable):

Title:

Phone Number:

E-mail Address:

3. Collective Bargaining Agent(s)

List information on each collective bargaining unit separately.

Union Name and Local Number:

Agent's Name:

Address:

Phone Number:

E-mail Address:

4. Number of Employees and Contractor Employees

Number of Employees Working at Applicant's site:

At time of submission of the VPP application, if the number of employees varies significantly during the year, mention that and include the average number of full-time and part-time employees with the range. Also include the reason for the variance, such as: energy demand and seasonal drivers such as in the postal service during the year-end holiday season and before Mother's Day.

Number of Temporary Employees supervised by Applicant:

Estimate of the number of temporary employees generally supervised at time of submission of the VPP application. Temporary employees are those employees that are working for a limited appointment but are directly supervised by the applicant. They may include students working during school breaks or other transient workers hired for a specific project. If the number of employees varies significantly during the year, mention that and include the average number of full-time and part-time employees with the range. Also include the reason for the variance, such as: energy demand and seasonal drivers such as in the postal service during the year-end holiday season and before Mother's Day, summer temporaries working during the school break.

Number of Applicable Contractor Employees:

A contractor or contract worker is one who is expected to perform a specific function under his or her own direction. They may receive general guidance but the method of work is determined by the contractor or contract worker. Typical contractors include the construction trades, housekeeping, grounds keepers, cafeteria, and computer management, and security. An applicable contractor is one whose employees have worked more than 1000 hours at the applicant's work site during any one calendar quarter in the previous calendar year. List the applicable contractors and the services provided.

5. Type of Work Performed and Products Produced

Provide a comprehensive description of the work performed at your site, the type of products produced, and the type of hazards typically associated with your industry.

6. Applicant's Standard Industrial Classification (SIC) Code and North American Industrial Classification System (NAICS) Code

Provide your site's SIC and NAICS codes.

Use the full four- and six-digit codes for the SIC and NAICS, respectively, for the primary product or service produced or provided by the work site.

7. Recordable Injury plus Illness Case Incidence Rates

Complete and submit the work site table at the end of this application (Section G). Then:

The tables and instructions are detailed later.

- *Record your combined 3-year total case incidence rate (TCIR) for recordable nonfatal injuries and illnesses here.*

  TCIR is the term used to refer to the total case incident rate. It is a calculation of the following columns of the OSHA 300 logs: H, I, J.

- *Record your days away, restricted, transferred incidence rate for recordable injury and illness cases involving days away from work, restricted work activity, and/or job transfer (DART) here.*

  DART is the term used to refer to the days away, restricted, transferred case incident rate. It is a calculation of the following columns of the OSHA 300 log: H, I.

The TCIR and DART rates are calculated as follows using the data on the OSHA 300 logs or the 300A summary of work-related injuries and illnesses.

The 3-year TCIR rates are calculated as follows: 200,000 × totals for columns H, I, J divided by total hours worked including temporary employees for each of the previous 3 calendar years.

The 3-year DART rates are calculated as follows: 200,000 × total for columns H, I divided by total hours worked including temporary employees for each of the previous 3 calendar years.

Note that the 3-year rates are not an average of the rates for the 3 previous years but a compilation of the data for all of the last 3 years.

7a. Rates at or above Industry Average

If, after completing Table 3.1, you determine that your 3-year TCIR, DART, or both are at or above your industry average for each of the last 3 years, specify your short- and long-term goals for reducing these rates to a level below the industry average. Include specific methods you will use to address this problem. It must be feasible to reduce rates within 2 years.

To be able to determine the comparison between the BLS average rates and the applicants 3-year rates, the BLS rates must be obtained. They can be obtained from the table of incidence rates[1] of nonfatal occupational injuries and illnesses by industry and case types on the BLS website: http://www.bls.gov/iif/oshsum.htm. The applicant's rates must be compared to the rates in the table for the appropriate NAICS code. Since the BLS rates are averages derived from data collected from a annual survey, there may not be enough data for the detailed six-digit NAICS code for a specific industry. One example of this is the 2007 data for NAICS code 114111 for finfish fishing; the closest data available from the BLS table is for 1141 for fishing. In that case, the comparison would be to the NAICS code 114100 for fishing.

The TCIR rates are compared to the BLS rates in the column headed Total Recordable Cases, and the DART rates are compared to the BLS rates in the column headed Cases with Days Away from Work, Job Transfer, or Restriction-Total.

Table 3.1 presents a sample of the table required in the application. Using this table, the calculations for the 3-year TCIR and DART rates would be calculated using the formula discussed previously:

$$\text{TCIR} = \frac{200{,}000 \times 42}{2{,}842{,}000} = 3.0$$

$$\text{DART} = \frac{200{,}000 \times 23}{2{,}842{,}000} = 1.6$$

To determine the percentage of difference between the work-site's rates and the BLS average rates, use the following formula: Site rate minus BLS rate/(BLS rate × 100).

TABLE 3.1 Table of Recordable Injuries and Illnesses

	A	B	C	D	E	F	G	H	I
Year	Total Work Hours	Total No. of Injuries	Total No. of Illnesses	Sum of Injuries and Illnesses	TCIR[2]	Total No. of DART Injury Cases	Total No. of DART Illness Cases	Sum of DART Injury and Illness Cases	DART[3] Rate
3 years ago (2006)	952,000	13	2	15	3.2	8	1	9	1.9
2 years ago (2007)	960,000	15	1	16	3.3	9	0	9	1.9
Last year (2008)	930,000	11	0	11	2.4	5	0	5	1.1
3-year totals and rates	2,842,000	39	3	42	3.0	22	1	23	1.6
2007 BLS rates for NAICS 123456					4.3				3.1
Percent below/above BLS national average[4]					−30%				−48%
Current year to 12/12	410,000	5	0	5	2.4	2	0	2	1.0

Using the data from the previous table, the comparisons of the TCIR and DART to the BLS data would be calculated as follows:

BLS data: TCIR = 4.3; DART = 3.1

3-year data: TCIR = 3.0; DART = 1.6

TCIR compared to BLS = $(3.0 - 4.3)/(4.3 \times 100) = -30\%$

DART compared to BLS = $(1.6 - 3.1)/(3.1 \times 100) = -48\%$

It must be noted that the feasibility of reducing the rates refers to both the mathematical and practical feasibility to reduce the rates. Mathematically, it would be almost impossible to reduce the rates if the most recent full calendar year rates are greater than three times the most recent BLS rates for the specific industry code. Even with zero incidents in the following two full calendar years, the company's 3-year rate would still likely be above the BLS rates, considering that the BLS rates have been in a mostly steady decline for most industries.

Practical considerations include such factors as: labor management issues, cost of required engineering controls, organizational dynamics affected by budget reductions, and so forth.

Whereas OSHA will accept a goal of reducing the injury and illness rates over 2 years, such a goal must be supported with objectives to achieve the goal. The objectives must meet the criteria of SMART objectives (specific, measurable, achievable, realistic, and timely).

Let us present an example: We-R-Safe has recognized that its TCIR rates are above the industry average because of a large number of recordable significant hearing threshold shifts among its workers. They have identified the prevalence in one specific area of the plant. To achieve a reduction in these incidents, the safety manager has developed specific objectives that include:

- A comprehensive noise survey of the entire plant.
- Studying the available engineering controls to reduce the noise (antivibration pads, noise dampers, enclosures, greater separation between equipment).
- Development of a hearing conservation program supported with employee training on the program.
- Employee involvement in hearing protection personal protective equipment. All employees would also receive new audiograms to develop a new baseline.

7b. Alternative Rate Calculation for Small Businesses

If you are a small business (fewer than 250 employees at the site and fewer than 500 employees corporate wide), you may be eligible for the alternative rate calculation. Contact your regional OSHA VPP manager or state plan VPP coordinator or review the VPP Federal Register notice of July 24, 2000 for more details.

Recognizing that smaller work sites may be more affected by only one or two recordable incidents as compared to larger work sites in the same industry, an alternative rate calculation was developed and is available to those defined as smaller work sites.

The alternative rate calculation, if the work site qualifies, allows the use of the best 3 out of the most recent 4 complete calendar years' injury and illness recordable incidence experience. The following criteria must be met to qualify for this benefit:

1. Using the most recent calendar year's hours worked, calculate a hypothetical TCIR assuming that the employer had two cases for the year.
2. Compare the hypothetical rate to the 3 most recently published years of BLS combined injury/illness total case incidence rates for the industry. This rate can be found in the table of "Incidence rates[1] of nonfatal occupational injuries and illnesses by industry and case types" on the BLS website (http://www.bls.gov/iif/oshsum.htm).
3. If the hypothetical rate is equal to or higher than the BLS rate in at least 1 of the 3 years, the employer qualifies for the alternative rate calculation method.
4. Table 3.2 illustrates the use of the alternative rate calculation for a roofing contractor with 15 employees.

In the example, since the hypothetical rate for two recordable incidents in one year is 13.3, which is greater that the BLS TCIR rate of 6.5, the applicant would be able to use the alternate rate calculation. That means that the highest year's rates (2007) of the most recent 4 calendar years would be removed and replaced with the lower 2005 data.

7c. Rates for Applicable Contractors

Complete Table 2 for rates of Applicable Contractors (those who worked more than 1000 on your site in any calendar quarter), listing each contractor individually. (This table does not get submitted with the application.)

B. Management Leadership and Employee Involvement
Management Leadership

1. Commitment

Attach a copy of your top-level safety policy specific to your facility. Note: Management must clearly demonstrate its commitment to meeting and maintaining the requirements of the VPP and taking ultimate responsibility for worker safety and health.

TABLE 3.2 Alternative Rate Calculation

Year	Hours	Total Cases	TCIR
2005	30,000	0	0.0
2006	30,000	1	6.6
2007	30,000	2	13.3
2008	30,000	1	6.6
3-year rate 2006–08	90,000	4	8.9
Alternate rate 2005, 06, 08	90,000	2	4.4
2007 BLS for NAICS 23816			6.5
Test	30,000	2	13.3

The policy should be written and signed by the site's chief executive officer (plant manager, general manager, etc.). The policy must be specific to the work site and must be shared with all employees. The evaluation team will ask the interviewees if they are aware of it and if they understand the principles contained in it.

2. Organization

Briefly describe how your company's safety and health function fits into your overall management organization. Attach a copy of your organization chart.

The organization chart should illustrate the relationship between the senior safety and health management official to the site's senior executive. There should also be an explanation if the safety and health senior official is a full-time position or if safety and health is a collateral duty of another position. It must be noted that having safety and health as a collateral duty is not in itself a negative. Other factors will be taken into consideration, such as size of the workforce and complexity of the workplace.

3. Authority and Responsibility

Describe what authority you give managers, supervisors, and regular employees regarding safety and health and hazard mitigation.

The more effective safety and health management systems empower all employees to at least stop those activities that represent potential exposure to recognized hazards. The description should detail how this is accomplished and how those hazards are mitigated. Explain any differences in the authority assigned to managers, supervisors, and employees.

4. Accountability

Briefly describe your accountability system used to hold managers, line supervisors, and employees responsible for safety and health. Examples are job performance evaluations, warning notices, and contract language. Describe system documentation.

This description should detail any differences between the accountability systems for salaried and hourly employees. It is recognized that union-represented employees do not generally receive formal performance evaluations, whereas salaried employees do receive them. The description must contain the specific safety and health goals and objectives for which salaried employees are responsible. One typical area of improvement is that the goals that are reviewed are very general, such as "support the safety and health policy" or "provide a safety and healthful workplace for the employees." The goals and objectives should be based on SMART objectives (specific, measurable, achievable, realistic, and timely), and they may vary depending on the position. Examples include: completion of training, participation in safety committee meetings, number of inspections participated in, participation of direct reports in safety and health activities, participation in job hazard analyses, and the like.

5. Resources

Identify the available safety and health resources. Describe the safety and health professional staff available, including appropriate use of certified safety professionals (CSP), certified industrial hygienists (CIH), other licensed health care professionals, and other experts as needed, based on the risks at your site. Identify any external resources (including corporate office and private consultants) used to help with your safety and health management system.

In addition to listing the staff resources, detail how they are used. Certified Industrial Hygienists may perform hazard assessment surveys and sampling strategies as well as perform personal and area monitoring. Professional licensed health care professionals may perform ergonomic assessments and recommend corrective actions to minimize body motion stressors; consultants may develop special hazard control programs or perform third-party audits of the safety and health management system. Also include how any corporate safety and health staff are used at the site. Additional resources for safety and health information and assistance may be obtained from the applicant's workers' compensation insurance carrier and outside safety and health consulting firms. The application should explain how the insurance carrier and consultants have been used in the past as safety and health resources. They may have performed audits and sampling, written programs, developed and provided training, or performed an annual evaluation of the entire safety and health management system.

6. Goals and Planning

Identify your annual plans that set specific safety and health goals and objectives. Describe how planning for safety and health fits into your overall management planning process.

This section should address both the planning process specific to the safety and health activities at the work site as well as how safety and health is considered in the overall business and operational planning process. Factors to consider in the safety and health planning include: scheduled sampling and monitoring based on the sampling strategy and plan, the development of the annual training plan, purchase of new or replacement personal protective equipment, scheduled preventive maintenance, and the like. Factors to consider for safety and health in the business plan include: expansions or contractions of the workforce, initiation of new production schedules or products, and revisions to the production process, proposed facility expansions or modifications, purchase and installation of new equipment, and so forth.

7. Self-Evaluation

Provide a copy of the most recent annual self-evaluation of your safety and health management system. Include assessments of the effectiveness of the VPP elements listed in these application guidelines, documentation of action items completed, and recommendations for improvement. Describe how you prepare and use the self-evaluation.

This is one of the most misunderstood requirements of the VPP application and process. The self-evaluation is not a compliance-focused review of the hazard control programs of the work site. Nor is it an audit or inspection of the work site to ensure that there are no uncontrolled hazards. Although it does contain both of these activities, it is much more. The self-evaluation is best described as a critical review of the effectiveness of each of the elements of the safety and health management system. It reviews the activities for the previous year and highlights those elements that have worked effectively and those that need improvement. This requirement will be discussed in greater detail later in this chapter. This section should contain a description of how the evaluation is performed. The evaluation can be completed by the safety and health staff, by the safety committee, or any other individual or group in the company. It may also be completed by a third party such as a consultant or the insurance company. The prime consideration is that whoever is performing the self-evaluation understands the safety and health program and the principles of the evaluation process. The most critical idea is to be totally objective. One of the greatest failures of this requirement is that those performing the self-evaluation are not absolutely objective, and they are reluctant to report unfavorable findings. This is not "Sleeping Beauty" and you are not the villian Queen asking the magic mirror "who is the fairest of them all?" and always expecting the answer to be you. You have to expect and accept that you will not always be the fairest of them all. Accept the blemishes as an opportunity to improve your safety and health management system.

Employee Involvement

8. Three Ways

List at least three meaningful ways employees are involved in your safety and health management system. These must be in addition to employee reporting of hazards. Provide specific information about decision processes in which employees participate, such as hazard assessment, inspections, safety and health training, and/or evaluation of the safety and health management system.

Include all those ways that employees are provided an opportunity to participate in the safety and health process. That may include:

- Committee members
- Various safety or health teams (confined space rescue, medical emergency response, hazardous materials response team, fire brigade, etc.)
- Inspections
- Job hazard analyses
- Developing training programs
- Behavioral safety observations
- Submitting safety suggestions

OSHA does not consider reporting unsafe conditions as an example of employee involvement since that is a right that all employees have.

9. Employee Notification

Describe how you notify employees about site participation in the VPP, their right to register a complaint with OSHA, and their right to obtain reports of inspections and accident investigations upon request. (Methods may include new employee orientation; intranet or email if all employees have access; bulletin boards; toolbox talks; or group meetings.)

First we must emphasize that this is a combination of an OSHA compliance issue and an administrative requirement of the VPP. OSHA should not be too critical if interviewed employees are not aware of these rights, although it will be a positive sign if they are. The most important thing to understand is that the requirement is that the employees have been informed of the site's participation in the VPP and their OSHA rights. To quote Peggy Richardson as she described it to one of the authors: "It is management's responsibility to teach the employees about the VPP; not to learn them." This section should detail how the employees were so notified. Remember that the most effective method of communication is direct conversation. Toolbox talks, group meetings, and other similar active activities are more effective than the passive activities such as e-mail and bulletin board postings.

10. Contract Workers' Safety

Describe the process used for selecting contractors to perform jobs at your site. Describe your documented oversight and management system for ensuring that all contract workers who do work at your site enjoy the same healthful working conditions and the same quality protection as your regular employees.

This section should detail the prequalification process you use to evaluate the ability of contractors to work safely at the work site. Describe the materials reviewed as submitted by the contractors. These may include:

- Experience modification rates (EMR) for workers' compensation insurance premiums including any cutoff such as a maximum EMR of .75, recent years' injury and illness rates
- OHSA inspection history
- Safety and health policy
- Hazard prevention programs
- Training program

Also describe how the work site ensures that the contractor work is not introducing hazards to the workplace. Examples may include: routine inspections of contractor activity, reviews of chemicals brought on to the site, ongoing training, and the like.

11. Site Map

Attach a site map or general layout.

Although the *Federal Register* requires that a site map be submitted with the application, many VPP coordinators suggest that this section contain an assurance that the site map will be available for review during the OSHA VPP onsite evaluation. This is because the entire application becomes part of the public record once the work site is approved as a VPP participant. Since the application is a public record, it may be requested by just about any individual, business, or group. Rather than expose any potential security information to the public, it has become the practice to overlook this requirement. However, the authors suggest that the local VPP coordinator be contacted to determine his or her position.

C. Work-Site Analysis

1. Baseline Hazard Analysis

Describe the methods you use for baseline hazard analysis to identify hazards associated with your specific work environment, for example, air contaminants, noise, or lead. Identify the safety and health professionals involved in the baseline assessment and subsequent needed surveys. Explain any sampling rationale and strategies for industrial hygiene surveys if required.

This is a multipart process that includes a qualitative assessment of the hazards that are or may be present in the workplace and the formal evaluation of those hazards using quantitative assessments such as air and noise sampling. The purpose of this element is to identify those concerns that warrant detailed sampling and monitoring so that potential exposures may be identified and appropriate control programs put into place to eliminate or reduce those exposures. This description should address all safety and health surveys that were performed to identify those hazards that are present in the workplace. The answer must also include a description of any sampling including the protocol and standards of reference, such as the OSHA permissible exposure limits (PELs), the American Conference of Governmental Industrial Hygienists (ACGIH) threshold limit values (TLVs), or National Institute of Safety and Health Recommended Exposure Limits (NIOSH RELs). The assessments must be performed by a qualified individual and should include both a physical tour of the workplace and a review of associated documentation, such as material safety data sheets for chemicals present in the workplace.

2. Hazard Analysis of Routine Jobs, Tasks, and Processes

Describe the system you use (when, how, who) for examination and analysis of safety and health hazards associated with routine tasks, jobs, processes, and/or phases. Provide some sample analyses and any forms used. You should base priorities for hazard analysis on historical evidence, perceived risks, complexity, and the frequency of jobs/tasks completed at your work site. In construction, the emphasis must be on special safety and health hazards of each craft and phase of work.

Different industries and companies use different terms to describe these analyses. Typical terms in general industry include: job safety analysis (JSA) and job hazard analysis (JHA). Chemical industry terms include: process hazard analysis (PHA) and hazardous operations reviews [HAZOP). The construction industry uses: activity hazard analysis (AHA), task hazard analysis (THA), and phase hazard analysis (PHA). These terms are not exclusive and there are others that are in use as well. Other industries may also have their own terminology. Regardless of the term used, the principles are the same. Each job, task, or process is reviewed to identify each step of the operation, the hazards associated with the step, and the appropriate protection recommended to eliminate or reduce the potential for injury or illness. Following is a sample of one of our job hazard analyses.

The description must include the method used; the qualifications and training of the individuals performing these analyses, if they are managers, employees, or consultants; how the analyses are used; how frequently they are reviewed (preferably at least annually); and the training employees receive on them. The analyses may also be integrated into standard operating procedures (SOPs), or the existing SOPs may be used to develop the analyses. A critical factor is the involvement of as many operational employees as possible.

3. Hazard Analysis of Significant Changes

Explain how, prior to activity or use, you analyze significant changes to identify uncontrolled hazards and the actions needed to eliminate or control these hazards. Significant changes may include nonroutine tasks and new processes, materials, equipment, and facilities.

The most important factor in this, as well as most other elements, is employee involvement. The application should emphasize how employees are involved in evaluating any changes to processes, equipment, and chemicals. The usual terms for this exercise are management of change (MOC) or new and altered equipment reviews. The employee involvement is expected to include both operators as well as maintenance staff. Operators are usually most familiar with the process. They and the maintenance staff have a vested interest in the design and operation of the equipment and are the groups very familiar with the requirements of the process. If the equipment cannot be easily operated and maintained, repairs will take longer and may entail more hazards than would be present if the maintenance staff were initially involved in the design and placement of the new equipment. One example of such a problem was the installation of a new large paper machine that did not involve the maintenance staff. The access areas for the mechanics to work on the machine were on the wall side of the equipment and the machine was installed too close to the wall thus inhibiting easy access. This resulted in postinstallation modifications to be able to facilitate maintenance and repairs. Not only does this add significantly to the cost, but it delayed the startup of the machine.

A positive example that has been seen on numerous occasions is involving the operators throughout the entire change process. They should be consulted about their expectations and concerns and encouraged to make suggestions. Some VPP companies have sent teams of management, operational, and maintenance employees to the equipment manufacturer to review the equipment and participate in the development of specifications. Some OSHA standards such as Process Safety Management specifically require employee involvement in the hazard analysis process.

4. Self-Inspections

Describe your work-site safety and health routine general inspection procedures. Indicate who performs inspections, their training, and how you track any hazards through to elimination or control. For routine health inspections, summarize the testing and analysis procedures used and qualifications of personnel who conduct them. Include forms used for self-inspections.

The Voluntary Protection Programs (VPP) require that inspections in general industry occur at least monthly, with the entire facility being inspected at least quarterly. Construction projects must have some inspection activity at least weekly, with the entire project inspected at least monthly. The greater frequency in construction is to address the constantly changing. Many larger facilities have addressed this requirement by using one of the following practices:

- Some divide the facility into defined sections that are each inspected monthly by designated teams thus ensuring the inspection of the entire facility monthly.
- Others use the same idea of sectioning the facility, but instead perform an inspection each month in one third of the facility. That ensures that the requirements of the VPP are being met. The description should describe the method used.

The description should emphasize the training of inspectors and the tracking of hazards found. OSHA is very interested in the training of inspectors and suggests a hazard recognition course for inspectors. We have also found that involving employees from other areas or departments on inspection teams gives them more appreciation of the safety and health management system, makes them more aware of their work environment, and provides a fresh set of eyes looking for unsafe conditions. Although there are several over-the-counter tracking programs (e.g., SAP, DBO2, MP2), Microsoft Excel is also very useful. The authors do not recommend one method over another but do recommend using some form of automated system. The most important elements of the system are details of the hazard, recommended corrective action, assigned completion date, and responsible person or position.

5. Employee Reports of Hazards

Describe how employees notify management of uncontrolled safety or health hazards. Explain procedures for follow-up and tracking corrections. An opportunity to use a written form to notify management about safety and health hazards must be part of your reporting system.

The VPP places a significant emphasis on employee involvement. However, since OSHA considers this a very basic factor in all safety systems, it does not recognize this as a method of employee involvement. Notification may be formal, allowing employees to directly enter hazard correction items into the work order system, or informal by simply informing their supervisor, foreman, or union steward. They should also be offered a means to report such uncontrolled safety or health hazards anonymously. With all of this, the most important factor is that the employees be aware of these procedures as well as the belief that such reports will be responded to. The second most important factor is the corrective action taken by management and the tracking of such actions.

6. Accident and Incident Investigations

Describe your written procedures for investigation of accidents, near misses, first-aid cases, and other incidents. What training do investigators receive? How do you determine which accidents or incidents warrant investigation? Incidents should include first-aid and near-miss cases. Describe how results are used.

Most facilities, VPP or not, investigate the incidents they are aware of. The authors have discussed this with numerous employees and managers and have concluded that many incidents are not reported. Fortunately, they are mostly of a minor nature with either no or just minor injuries that are self-treated. The concern is that these minor incidents may represent an unrecognized trend that is not being captured and eventually they may result in a more serious injury. The application should contain an explanation that addresses how the company encourages all employees to report all incidents, no matter how minor. One way to ensure this happens is to balance a strong disciplinary system for failure to report with positive encouragement such as minor recognition for reporting.

Describe how once the investigation is complete the investigation results are used to create corrective actions and revised procedures to avoid recurrences. Corrective actions must be assigned a due date and tracked to completion, which should be explained in the application. Inspection procedures should also be revised based on the incident. The application must contain a copy of the accident investigation forms, including sanitized copies of a few completed forms.

7. Pattern Analysis

Describe the system you use for safety and health data analysis. Indicate how you collect and analyze data from all sources, including injuries, illnesses, near-misses, first-aid cases, work order forms, incident investigations, inspections, and self-audits. Describe how results are used.

Trend analysis may range from simply identifying those items that have occurred frequently to detailing the trends by item, body part, shift, time of day, day of the week, day of the month, area of the plant, and other related factors. Clearly, the more detailed analyses would not work effectively for many smaller companies with fewer incidents, but they are very effective for larger companies. Including the leading indicators such as those above as well as training, employee involvement, and behavioral observations enables more data points to the analysis. The application should include a list of those items trended and how the trends are used to focus attention on the causative factors.

D. Hazard Prevention and Control

1. Engineering Controls

Describe and provide examples of engineering controls you have implemented that either eliminated or limited hazards by reducing their severity, their likelihood of occurrence, or both. Engineering controls include, for example, reduction in pressure or amount of hazardous material, substitution of less hazardous material, reduction of noise produced, fail-safe design, leak before burst, fault tolerance/redundancy, and ergonomic design changes.

It is typical for VPP application authors to not recognize the engineering controls in place at their work sites. These controls have been designed into the process and have become routine. The exceptions would include those that have been instituted recently. For example, exposure to significant occupational noise may be controlled by either setting equipment apart a suitable distance or using vibration insulators or enclosures. However, these may not be recognized because that is the way the original equipment was designed and installed. On the other hand, if the equipment was modified recently to address previous high noise exposures that would likely be recognized as an engineering control. Engineering controls should be described regardless of when they were implemented, either in the original design or as a modification.

Although not as reliable as true engineering controls, this category also includes protective safety devices such as guards, barriers, interlocks, grounding and bonding systems, and pressure relief valves to keep pressure within a safe limit.

The application should also contain a description of other engineering controls present in the workplace to prevent employee exposure to mostly physical contact with the hazards. Examples include:

- Light curtains
- Pressure-sensitive mats
- Magnetic interrupt switches
- "Dead man" controls

2. Administrative Controls

Briefly describe the ways you limit daily exposure to hazards by adjusting work schedules or work tasks, for example, job rotation.

The application must address administrative controls in all areas of the workplace. These include production, shipping and receiving, warehousing, and all administrative areas. Typically, administrative controls are considered to eliminate or mitigate ergonomic hazards. These start with performing ergonomic assessments and training employees to understand the hazards. Ergonomic hazards are limited in the administrative areas with such techniques as focus on correct office workplace design including chairs and desks, keyboard, mouse and monitor placement, ambient noise levels, lighting, and the like. In the materials handling areas these hazards may be limited by the use of powered moving and lifting equipment, vacuum systems, proper design of conveyor systems, placement of materials, and so forth. In the production areas, consideration must be given to the work environment and physical stressors of the job. Production area administrative controls may include job rotation, scheduled work breaks, hydration, cooling or heating stations, weight limitations for manual lifting, spring-loaded tables, and the like.

3. Work Practice Controls

Describe and provide examples of your work practice controls. These include, for example, workplace rules, safe and healthful work practices, specific programs to address OSHA standards, and procedures for specific operations that require permits, labeling, and documentation. Identify major technical programs and regulations that pertain to your site, such as lockout/tagout, process safety management, hazard communication, machine guarding, and fall protection.

With the exception of the Respiratory Protection Program, which is discussed in the following section on personal protective equipment, most OSHA VPP coordinators do not require copies of the entire individual work practice control programs. At a minimum, the application must list all such programs. This can be accomplished with the submission of a copy of the table of contents of the Safety and Health Manual. Each of the significant programs can be discussed in summary fashion and supported with copies of both blank and completed permits. Examples of permits to be included in the application include:

- Lockout/tagout (LOTO)
- Confined space entry
- Hot work
- Powered industrial trucks
- Lifting equipment
- Scaffolds

Other non-compliance-related work practice control programs should be briefly discussed. These may include workplace rules such as standard operating procedures, ergonomics, permit processes, and other factors that are not discussed elsewhere in the application.

4. Personal Protective Equipment

Describe and provide examples of required personal protective equipment (PPE) your employees use and what PPE the OSHA team members will need to bring to your work site.

Include in this description a list of the different types of PPE in use at the work site. Although many OSHA VPP coordinators do not require the submission of the complete Respiratory Protection Program, the *Federal Register* does require it. This issue must be referred to the OSHA VPP Manager or VPP Coordinator to determine their expectations. The standard practice is to accept a summary of the program rather than the entire program. Usually, it is sufficient to submit a copy of the table of contents. In addition, indicate if voluntary use of filtering face piece respirators (N95 dust masks) is allowed and how employees are trained on their use and limitations. Appendix D of the OSHA Respiratory Protection Program standard (29 CFR 1910.134) must be provided to such employees. Provide a description of the training and enforcement programs for PPE use. Also include a discussion of how PPE is selected, considering the hazard assessment and employee involvement in the selection process.

5. Enforcement of Safety and Health Rules

Describe the procedures you use for disciplinary action or reorientation of managers, supervisors, and other employees who break or disregard safety and health rules.

Regardless of the commitment of all employees to the safety and health management system and the safety and health rules, all such rules must be supported by an effective disciplinary procedure. OSHA's primary concerns include the principle that all levels of employees are covered either by the same or equivalent disciplinary programs, and that the programs are used equally without discrimination. Another very important factor for all such programs is that the employees must know about them. Unless the employees know the negative consequences for violating safety rules, they may be tempted to do so.

Most disciplinary programs follow a progression of violations to add to the severity of the sanction. These programs also allow for the progression to be advanced based on the severity of the infraction. One example of a sanction that the authors found to be very effective was requiring the affected employee to discuss their infraction and its potential safety or health consequences with other employees to dissuade them from doing the same thing. This is most effective if the affected employee can actually deliver an effective message. The most significant benefit of this is the delivery of the safety or health message by a fellow employee instead of a manager or trainer.

6. Preventive/Predictive Maintenance

Summarize your written system for monitoring and maintaining workplace equipment to predict and prevent equipment breakdowns that may cause hazards. Provide a brief summary of the type of equipment covered.

The primary concerns that should be included in the description regarding the preventive maintenance of equipment are:

- Protocols
- Training
- Scheduling
- Meeting schedules

Describe how the manufacturers' recommendations are addressed to maintain equipment or if equipment is maintained on another basis based on experience. Describe how employees are trained on the maintenance procedures, as well as applicable safety rules such as energy control, chemical awareness, and permits. One very effective system integrates the safety systems into the preventive maintenance work order system. The application should include a discussion of the scheduling system to ensure that preventive maintenance is completed in accordance with the requirements. There are many over-the-counter software scheduling programs, and many companies develop their own such programs. Either method is satisfactory so long as it works effectively.

Many companies also use a process of predictive maintenance using techniques such as vibration and lubrication analysis to determine the effectiveness of the equipment. This is not a discussion of the principles of predictive maintenance but it is addressed to ensure that the application does describe such a process used by the applicant.

The application should include the types of equipment covered by the preventive maintenance procedures. These would include operational equipment, as well as other supporting equipment such as hoists, cranes, rigging, maintenance, lifts, material handling, emergency response, and the like.

7. Occupational Health Care Program

Describe your onsite and offsite medical service and physician availability. Explain how you utilize the services of licensed occupational health care professionals. Indicate the coverage provided by employees trained in first aid, CPR, and other paramedical skills, their training, and available equipment.

The following questions must be answered in this section: Are health care professionals present at the work site and when? Are they used to evaluate workplace hazards and make recommendations for corrective programs? Are they used to perform fit-for-duty medical evaluations? Are they used to review medical surveillance documents such as those required by the OSHA respiratory protection standard? Are they involved in return-to-work programs? Describe how they are used for each of these concerns.

Describe how medical emergencies at the plant are addressed. Many companies have a medical facility that provides this service with support from an employee medical response team. Other companies rely exclusively on the employee team. These teams are usually voluntary and may consist of dedicated responders that have either a duty to respond or respond as Good Samaritans. Regardless of the level of obligation to respond, explain the level of training the team receives and their capability to provide first aid care and CPR if necessary. Detail the equipment they are supplied with, especially if they have an automated electronic defibrillator (AED). Also explain how they are notified in case of need.

If you use outside responders to provide medical support, explain how they have become familiar with the workplace and its hazards, how often they practice with the company responders, and what is their response time to the site.

8. Emergency Preparedness

Describe your emergency planning and preparedness system. Provide information on emergency drills and training, including evacuations.

The description must include the types of emergencies addressed in the emergency response plan such as fire, explosion, bomb threat, chemical release, employee or other violence, and natural disasters. Explain how employees are informed of such emergencies and their expected response actions. For example, are they expected to mitigate the emergency or evacuate the area? For those employees expected to mitigate the emergency, describe the specific training provided and the level and frequency of the training. This training may include the use of fire extinguishers and hazardous waste operations and emergency response (HAZWOPER) procedures.

The VPP application must describe how employees practice evacuations of the workplace and the frequency of the evacuation drills. The drills must be repeated for all shifts. Explain how those employees who miss a drill get to practice an emergency escape. Also explain how contractors and other visitors are integrated into the process. With regard to the drills, explain if company and outside responders are involved in the exercise. A very effective practice is to include a medical or other emergency personnel in the routine evacuation drill.

The expectations for construction projects differ from general industry because of the constantly changing workplace and environment. Employees at construction projects must be kept informed of such changes as soon as possible. Describe how they are kept informed of the emergency response procedures. This is usually addressed during the daily "toolbox talks" or other daily planning meetings.

E. Safety and Health Training

Describe the formal and informal safety and health training provided for managers, supervisors, and employees. Identify training protocols, schedules, and information provided to supervisors and employees on programs such as hazard communication, personal protective equipment, and handling of emergency situations. Describe how you verify the effectiveness of the training given.

OSHA personnel will look at the following elements of the training program when they visit the work site to evaluate the validity of the application:

- Training subjects
- Schedule
- Verification of training completed
- Testing
- Training curricula

Therefore, the VPP application must address these factors in sufficient detail to provide OSHA with the information requested. Explain how training programs are developed relative to the work process and the workplace hazards. Emphasize if employees and other subject matter experts are used to develop training programs or if they are purchased from outside sources. Also describe any use of computer-based training programs and if they are generic or tailored to the workplace.

The training schedule must be described to explain how it is developed to ensure that all required training is provided each year. Of critical importance is the method used to ensure that all required training is delivered and that any missed training is made up. Also of critical importance is how you verify that the subject matter is understood. Describe any written, computer, or practical evaluation tools that are used and what happens if an employee does not demonstrate adequate understanding of the subject. We have intentionally not yet distinguished between managers, supervisors, and employees in this section. However, there are significant differences in some of the training needs among these groups. Supervisors usually need the same operational safety and health training as other employees. They also should have training specific to supervisory and leadership skills.

A sample of a completed VPP application for general industry is in Exhibit 3.2.

REFERENCE

1. US Department of Labor, Bureau of Labor Statistics, Industry Illness and Injury Data, Summary Tables.

Exhibit 3.1 VPP Elements

Management Leadership and Employee Involvement

1. Commitment to Safety and Health Protection
    a. Policy
    b. Goals and Objectives
2. Commitment to VPP Participation
3. Planning
4. Safety and Health Program
5. Management Leadership
    a. Clear lines of communication
    b. Setting a positive behavior
    c. Open door policy
    d. Equal quality protections for all
    e. Defined safety and health responsibilities for all
    f. Authority commensurate with responsibilities
    g. Adequate resources
    h. Accountability for responsibilities
6. Employee Involvement
7. Contract Worker Coverage
    a. Safety and health management system
    b. Selection and oversight
8. Self-Evaluation of the Safety and Health Management System
    a. An annual written narrative report with recommendations
    b. Addressing all VPP elements and sub-elements
    c. Follow any format recommended by OSHA.
    d. In construction, the evaluation must be conducted annually and immediately prior to completion of construction

Worksite Analysis

1. Comprehensive Safety and Health Surveys
    a. Baseline survey
    b. Industrial hygiene sampling rationale and strategy
    c. Nationally recognized procedures
2. Hazard reviews of routine jobs, tasks, and processes
3. Hazard analysis of significant changes
4. Self-Inspections
5. Employee Hazard Notification
6. Accident/Incident Investigations
7. Trend Analysis

Hazard Prevention and Control

1. General Requirements
    a. Understood and followed by all affected parties;
    b. Appropriate to the hazards of the site;
    c. Equitably enforced
    d. Written, implemented, and updated by management as needed, and must
    e. Incorporated in training, positive reinforcement, and correction programs;
2. System of Hazard Controls
    a. Tracking hazard elimination or control in a timely manner
    b. Monitoring and maintenance of workplace equipment such as preventive and predictive maintenance
    c. An occupational health care program
3. Hierarchy of Hazard Controls
    a. Engineering controls
    b. Administrative controls

c. Work practice controls

d. Personal Protective Equipment

4. Process Safety Management

Safety and Health Training

1. Management Understanding of Responsibilities
2. All Employees Taught How to Recognize Hazards
3. All Employees Taught Safe Work Procedures
4. All Those Onsite Understand Emergency Procedures
5. All Employees Understand Use of PPE

Exhibit 3.2 Sample VPP Application

We-R-Safe Incorporated

OSHA Voluntary Protection Programs Application

Anytown, USA

January 2009

A. General Information

1. Applicant

Site Name: We-R-Safe Incorporated

Site Address: Anytown, USA

Site Manager: Brian Smith

Title: Plant Manager

Site VPP Contact for OSHA correspondence: Norman Lynch

Title: Safety and Health Manager

Phone Number: 555-123-6778

E-Mail Address: norman@wersafe.com

2. Company/Corporate Name

Name (if different from above):

Address:

VPP Contact (if applicable);

Title:

Phone Number:

E-mail Address:

3. Collective Bargaining Agent(s) (list information on each separately)

Union Name and Local #: N/A

Agent's Name:

Address:

Phone Number:

4. Number of Employees and Contractor[1] Employees

Number of Employees working at Applicant's site: 480

Number of Temporary Employees supervised by Applicant: 100–150

Number of Applicable Contractor* Employees: 35

We-R-Safe routinely utilizes one temporary personnel agency. We have temporary employees working for us at all times, ranging from 100 to 150, depending on business needs. These personnel are supervised by WE-R-SAFE and are, therefore, placed on our OSHA Log when injured, and summed into our total headcount and man-hours records.

Contract personnel are used on a regular basis, and referred to as Independent Contractors. These personnel are supervised by the contractor/owner. We utilize full time computer repair contractors, furniture installers, electrical, building maintenance contractors, lab services, cafeteria services, custodial services, grounds and pest control services, and security contractors. Although reviewed by us, recordable injuries and illnesses for the independent contractor employees are entered onto the contractor's OSHA Log.

The injury/illness rates for each of the applicable contractors will be available for review by the OSHA evaluation team.

5. Type of Work Performed and Products Produced

Provide a comprehensive description of the work performed at your site, the type of products produced, and the type of hazards typically associated with your industry.

We-R-Safe manufactures widgets for use in various industries including, automotive, aeronautics, machine manufacturing, and other manufacturing end users. Our manufacturing facility is divided into four Units of Production that focus on each of our primary markets. Our production lines include both automated processes for continuous run lines and more manually controlled processes for special order product lines. All of our production equipment is equipped with machine guards to protect employees from all moving parts. All guards are installed with electronic sensors to ensure that the equipment will not operate unless the guards are securely in place. We do not use any armful chemical in any of our processes and all materials are delivered by truck. Our finished products are also shipped by truck, either the customers or a delivery service.

In addition to the production facility, we have the following areas: receiving, warehouse, maintenance, office, quality assurance laboratory, cafeteria, health clinic, and a wellness center.

Our property is 100% non-smoking.

6. Applicant's Industrial Classification Codes

Provide what you believe to be your site's 6-digit North American Industry Classification System (NAICS) code and your 4-digit Standard Industrial Classification (SIC) code. Please contact your Regional VPP Manager or Coordinator if you are having difficulty identifying an appropriate code. You can also find NAICS and SIC information on the Bureau of Labor Statistics' website, www.bls.gov. OSHA ultimately will assign you a NAICS code for purposes of VPP.

The applicable NAICS code is 123456

The applicable SIC code is 7890

7. Recordable Nonfatal Injury and Illness

Using information from your OSHA injury and illness logs (OSHA-300), complete and submit Table 1 in Section G at the end of this application. Then:

- *Record your combined 3-year TCIR[2] here:* 3.0
- *Record your combined 3-year DART[3] rate here:* 1.6

B. Management Leadership and Employee Involvement

Management Leadership

1. Commitment

Attach a copy of your top-level safety policy specific to your facility. Note: Management must clearly demonstrate its commitment to meeting and maintaining the requirements of the VPP and taking ultimate responsibility for worker safety and health.

WE-R-SAFE is committed to a policy of "beyond compliance" which includes written standards for the safety and health of our associates by working towards achieving our vision of an injury free work place. The policy specifies goals and objectives of the safety and accident prevention program. This policy is renewed regularly and signed by the Company Group Chairman. It appears in our Site Safety Manual, Associates' Orientation Booklet and is posted at all sites.

The WE-R-SAFE Performance Metrics Team that is composed of senior managers including the Safety and Health Manager establishes the company goals each year. These goals are negotiated with each internal organization. Company goals are communicated on a quarterly basis to all employees via the President's review sessions and Company Goals Score Cards.

Top management is actively involved in the safety and health management system through WE-R-SAFE's Safety Champions. In addition, various incentives, involvement and reward & recognition programs have been in place since 1996. (See attachment 1: A copy of the signed site safety policy and vision).

2. Organization

Briefly describe how your company's safety and health function fits into your overall management organization. Attach a copy of your organization chart.

The Safety and Health Organization (SHO) is responsible for identifying applicable requirements and implementing programs to ensure continued improvement as well as maintaining an effective and highly visible accident prevention program. The SHO is also responsible for providing direction to all managers, supervisors and employees to assist them in participating in our safety and health activities. This function reports directly to the Plant Manager.
(See attachment 2: Organization Chart.)

3. Authority and Responsibility

Describe what authority you give managers, supervisors, and regular employees regarding safety and health and hazard mitigation.

The company is dedicated to the safety and health of all associates. WE-R-SAFE works towards the corporate vision of having an injury-free work place. The site safety policy holds the managers responsible for ensuring that all SHO requirements are met, for maintaining a safe and healthy work environment, and for enforcing safe work habits. The safety related roles and responsibilities of our organization are detailed in our Safety Roles and Responsibilities document which is attached. Some examples of responsibilities are:

Managers:

- Provide overall leadership for the safety process.
- Support the company's safety policy, giving safety equal emphasis with other functions such as production, cost and quality.
- Provide resources (time, money) necessary to maintain the safety process.
- Evaluate staff member's safety performance and insures that Safety is a part of each one's objectives, as appropriate.
- Ensure a safety structure is in place in the form of working committees and/or departmental coordinators.
- Meet on a periodic basis with the management and supervisory staff to review departmental safety activities, objectives, safety results and progress of outstanding safety items and concerns and methods undertaken to prevent recurrence.
- Ensure inspections are completed and that corrective actions are done in a timely manner. Will participate in the inspections a minimum of six times a year.
- Review accident reports and investigations of accidents, insuring corrective action is taken.

- Insure communication of safety matters to all levels of their respective organizations.
- Approve Safety Rules and Regulations on an as needed basis.
- Participate in the review of all serious injuries/illnesses.

Supervisors:

- Actively support the company's safety policy, giving safety equal emphasis with other functions such as production, cost and quality.
- Ensure the premises and equipment are in accordance with established safety standards.
- Act upon noted safety observations and initiates corrective actions.
- Include safety as part of SOPs and implements safety directives and guidelines.
- Participate in behavioral and safety inspections.
- Ensure all associates are instructed on safe job procedures and safety specific subjects. Ensure that such training is documented for associates' records.
- Ensure training is given to new or transferred associates.
- Ensure safety is part of the department meetings and attends the meetings.
- In case of accident, injury, illness or "near miss", makes certain that the associate receives immediate medical attention. Ensure a thorough investigation to determine cause(s) and corrective action(s). Complete the incident report within 24 hours.
- Evaluate employee compliance and enforce safety procedures and rules.
- Evaluate associate safety performance and insures that safety is a part of each associate's objectives.

The Site Executive Team provides overall leadership and resources. They participate in the review of all serious injuries/illnesses.

Safety Champions have been recruited to oversee the compliance of the various elements of the safety program. They assist in the development and administration of adequate safety systems and organizations by informing all departments of required safety structure and systems. They assist in performing safety audits.

Safety goals are built into each department's business plans. Management is responsible for the development and direction of the safety program and the Safety and Health Organization. Each business or functional manager, supervisor and associate has the responsibility to ensure that safety is part of the job.

(See Attachment 3: Safety Roles and Responsibilities.)

4. Accountability

Briefly describe your accountability system used to hold managers, line supervisors, and employees responsible for safety and health. Examples are job performance evaluations, warning notices, and contract language. Describe system documentation.

Department and operations managers use several methods for accountability for safety, including using the Performance Management Process (PMP). PMP includes performance appraisals for all associates based on specific goals and commitments. Safety and health goals and expectations are reviewed at all levels, as well as addressed in human resources policies. Site safety and health goals are shared in Management Review sessions with top level managers each quarter.

There are frequent E-mail messages showing positive feedback to all team members when safety walkthroughs show no issues. Site safety and health performance metrics are published on internal web sites, reviewed in the monthly safety representative meetings, in the site executive meetings, and presented in President's Quarterly Review sessions with all employees. We have a corporate-wide formal Quality Systems Policy that addresses multiple SHO issues along with product and service quality.

Our S&H Compliance Plan outlines commitment, organization, responsibility, accountability, resources and planning. Responsibility for safety is given to all associates and it is part of their performance plans. Accountability is integrated into our annual Corporate SHO self-audits and Corrective Action Plans (CAP) are generated. Manufacturing Standard Operating Procedures (SOPs) specify safe work and process procedures and personal protective equipment to utilize when performing jobs. Safety meetings are held at least monthly. Non-compliance with WE-R-SAFE policies and procedures may result in appropriate disciplinary action.

(See attachment 4: WE-R-SAFE Performance Review Form.)

5. Resources

Identify the available safety and health resources. Describe the safety and health professional staff available, including appropriate use of certified safety professionals (CSP), certified industrial hygienists (CIH), other licensed health care professionals, and other experts as needed, based on the risks at your site. Identify any external resources (including corporate office and private consultants) used to help with your safety and health management system.

The SHO team consists of a S&H Manager (CSP) with 18 plus years of experience, a staff industrial hygiene specialist (CSP, CIH) with 20 plus years of experience, an occupational safety specialist, an occupational health nurse (RN) with 20 plus years of experience, two health and fitness specialists, an employee assistance program consultant, plus additional support staff. The team uses a variety of references and resources to maintain current information for ensuring continuing compliance with S&H requirements and continuing performance improvement. Our corporate safety and health department supplies us, on a quarterly basis, with a CD-ROM with current government regulations, corporate safety manuals, and MSDSs.

Ongoing professional development takes place through American Society of Safety Engineers (ASSE) and American Industrial Hygiene Association (AIHA) professional development conferences, corporate steering committees and standards committees, corporate Technical Resources Group (TRG), and web sites access, etc. Multiple trade publications on safety, health, and industrial hygiene state-of-the-art and regulatory updates are maintained. Each year training and career planning and development sessions take place for all team members, as with all site associates. The site maintains a variety of safety and health monitoring equipment such as gas and air monitors, noise dosimeters, and various other meters.

6. Goals and Planning

Identify your annual plans that set specific safety and health goals and objectives. Describe how planning for safety and health fits into your overall management planning process.

The underlying goal of this worksite is 100% safe production. Prevention of all injuries and incidents is our continuing goal and is the driving force of all of our planning and activities. Specific annual goals are developed that address incremental improvements in our injury and illness rates and the level of employee involvement in our safety and health activities. Safety and incident prevention is in our culture and is considered part of the job. Safety and health is included in the annual business plan. Planning for safety and health fits into the overall management process, as members of the SHO team are included in staff meetings of manufacturing operations managers, facilities and engineering teams. As a result, project needs are identified and partnered. Regular and ongoing reviews of factory moves or construction projects are held. Safety performance metrics are presented in quarterly communications meetings with all associates, monthly safety representatives meetings, and in Site Management Review sessions with high level management at least quarterly. Required training in safety and health courses is published and mandated as required.

7. Self-Evaluation

Provide a copy of the most recent annual self-evaluation of your safety and health management system. Include assessments of the effectiveness of the VPP elements listed in these application guidelines, documentation of action items completed, and recommendations for improvement. Describe how you prepare and use the self-evaluation.

We complete a comprehensive evaluation of our entire safety and health management system every twelve months. This process starts in January and results in a formal report that must be completed by February 15. The Safety

Manager completes the evaluation with the collaboration of the Plant and Department Managers and Supervisors. The report is shared with the associates at monthly safety meetings, quarterly associate meetings, and the annual meeting with the Plant Manager. The report contains identified opportunities for improvement with recommendations for correction.

In addition to the annual self-evaluation we are also audited every three years by a team that represents the corporate safety and health organization. Those audits also follow the VPP recommended process and are done by selected Safety and Health Managers from other company plants.

(See attachment 5: Previous year's annual evaluation.)

Employee Involvement

8. Three Ways

List at least three meaningful ways employees are involved in your safety and health management system. These must be in addition to employee reporting of hazards. Provide specific information about decision processes in which employees participate, such as hazard assessment, inspections, safety and health training, and/or evaluation of the safety and health management system.

Employees are encouraged to identify and communicate opportunities for enhancing the safety and health of the workplace verbally, by telephone or through a variety of on-site programs. These include joining area safety teams, participating in safety inspections, inspectors asking associates where the next incident may occur during safety inspections, assisting in job hazard analysis, submitting the "Opportunity Improvement Form" for safety observations, near-miss reports, and ergonomic issues. The company launched a "Do It Right" initiative in January, 2000. This initiative allows associates to report issues on a hot line either openly or anonymously. All calls are received by an independent agency.

The site has various voluntary teams which focus on SHO initiatives such as Area Safety Teams, Equipment Manufacturing Safety Teams, Emergency Evacuation teams, Medical Emergency Response Team (MERT), Bio-Incident Response Team, Office Safety Team, etc. The teams consist of a number of identified or voluntary associates from the work areas and focus on injury/illness prevention, hazard evaluation, ergonomics and various new SHO programs and ideas, including VPP. Associates are invited to take part in incident investigations. Associates are empowered and encouraged to notify their supervisor and/or call the SHO at any time to report a hazard or take part in a job safety or ergonomics analysis.

There are charts posted in various locations that show our site injury/illness and lost work day rates each month, compared to our goals, and a break out of type of incidents.

S&H Days have been held three times per year. Information "booths" and displays are held for all associates. Dry erase boards have been set up in strategic areas such as for associates to report various issues including safety and health.

WE-R-SAFE Safety and Health Manuals and site safety information is available to associates on the WE-R-SAFE Intranet. There is also a Corporate CD-ROM available on the computer network that provides all associates access to corporate guidelines, safety manuals and current OSHA regulations.

(Information on the Do It Right Initiative, Safety & Environmental Recognition Program, Improvement Opportunity Form are available for review on site.)

9. Employee Notification

Describe how you notify employees about site participation in the VPP, their right to register a complaint with OSHA, and their right to obtain reports of inspections and accident investigations upon request. (Methods may include new employee orientation; intranet or email if all employees have access; bulletin boards; toolbox talks; or group meetings.)

Associates have been notified of the VPP program rights, responsibilities and process through a number of communication means. It was first introduced in two VPP presentations given by the Compliance Assistance Specialist of the local OSHA office. The presentation was also posted on our VPP web site and all associates were invited to view it. We had a poster campaign notifying associates on the benefits of the VPP Program. The OSHA Rights and Responsibilities are posted on associate bulletin boards. Associates have been empowered and involved in safety assessments, walkthroughs, and hazard reporting and incident investigations. VPP is introduced to all associates during the new-employee orientation. OSHA "It's The Law" Posters are posted at all employee entrances.

The availability of safety inspection and incident investigation results is presented in the new employee safety training.

All temporary employees, contractor employees and contractor management have also been notified of our drive toward participation in OSHA's VPP.

(Web Site Documents will be available for review on site using the We-R-Safe intranet)

10. Contract Workers' Safety

Describe the process used for selecting contractors to perform jobs at your site. Describe your documented oversight and management system for ensuring that all contract workers who do work at your site enjoy the same healthful working conditions and the same quality protection as your regular employees.

WE-R-SAFE utilizes one primary agency for temporary employees that are supervised by WE-R-SAFE managers. These temps receive the same safety and health services as do WE-R-SAFE employees. We ensure that they receive the new employee orientation within the first 30 days and are included in other S&H training and functions. Temps are identified with picture badges that are a different color than regular WE-R-SAFE personnel. They are provided with safety glasses, safety shoes and any other PPE or medical surveillance that may be required. They take part in S&H promotional events such as safety fairs and drawings, as well.

WE-R-SAFE manages independent contractors, who are on-site occasionally to perform work, using a Contractor Safety Program. Independent contractors also include all construction related contractors. This program requires those who hire contractors to use firms which have been pre-qualified on the basis of contractor safety and health records and past safety performance. Contracts are written to permit termination if a contractor fails to meet safety and/or health guidelines.

The Safety and Health Organization maintains a database of contract firms who are approved to work at WE-R-SAFE. These firms have demonstrated their regard for safety by maintaining below-average experience modification rates (EMRs) and incident statistics. New contract firms are considered for inclusion in the database, following a pre-qualification process. The firm must demonstrate their ability to meet or exceed safety, health and construction requirements. This may include providing written safety and health programs, policies and proof of training.

Periodic performance audits of independent contractors are also conducted. Contractor employees found to be out of compliance are stopped from continuing work and may be asked to leave the premises until a meeting with the contractor supervisor and location management can take place. An annual review of contractor safety performance is a determining factor in the continuing use of the contractor in subsequent years.

Prior to performing work on site independent contractors are required to have a site orientation, complete an outside contractor orientation checklist, and are issued a safety & health rules booklet. Following the orientation, contractors are issued special badges that identify them as outside contract personnel.

(Contractor Safety Policy and Procedures, Outside Contractor Orientation Checklist, Contractors Safety Guide, Contractor Performance Review, Film Operations Safety & Environmental Orientation, and Purchasing SOP – Identifying, Evaluating and Approving Suppliers are available for review on site).

11. Site Map

The site map will be provided to the OSHA Onsite Evaluation Team during their visit to the plant.

C. Worksite Analysis

1. Baseline Hazard Analysis

Describe the methods you use for baseline hazard analysis to identify hazards associated with your specific work environment, for example, air contaminants, noise, or lead. Identify the safety and health professionals involved in the baseline assessment and subsequent needed surveys. Explain any sampling rationale and strategies for industrial hygiene surveys if required.

A comprehensive facility-wide qualitative exposure assessment was completed in 2001 by a contract professional safety and health consulting company. Our qualitative exposure assessment was developed to identify all potential safety and industrial hygiene related hazards, including the often overlooked exposures from biologicals (like mold), light (both UV and infrared), heat and cold, and low frequency vibration. Results from the qualitative exposure assessment helped us to develop a sampling plan for more detailed quantitative sampling. Our SHO has managed the sampling plan. That has allowed us to identify which employees have the potential for over exposures. These employees have been included in appropriate medical surveillance programs. The facility's New Product Approval Policy may also trigger the need to update the qualitative exposure assessment as well as quantitative sampling.

We recognize the hazards associated with exposure to noise, and emphasize the use of hearing protection to protect employees from significant noise exposures at various areas of our facility. We provide hearing protection in the form of muffs and earplugs based on observations and informal interviews. All high noise areas have appropriate signage, and hearing protection equipment is readily available and widely used by site employees and contractors. Periodic audiograms are completed for employees, with the results reviewed by an Occupational Health Professional, and results are provided to the employees.

2. Hazard Analysis of Routine Jobs, Tasks, and Processes

Describe the system you use (when, how, who) for examination and analysis of safety and health hazards associated with routine tasks, jobs, processes, and/or phases. Provide some sample analyses and any forms used. You should base priorities for hazard analysis on historical evidence, perceived risks, complexity, and the frequency of jobs/tasks completed at your worksite. In construction, the emphasis must be on special safety and health hazards of each craft and phase of work.

WE-R-SAFE maintains an active program utilizing a Job Hazard Analysis (JHA) system. JHAs are required for all new processes and/or job tasks, whenever there is an incident (accident, injury or near miss) or for any reported safety or health improvement opportunity. The supervisor and safety staff work together to complete the JHAs for jobs or tasks in his/her area. Supervisors receive JHA training and use a JHA form available on the WE-R-SAFE Intranet to document the results of each JHA. Other options to JHAs are Standard Operating Procedures (SOPs) or Work Instructions that have been written to include safety procedures (such as personal protective equipment, work practices, use of ventilation, etc.). The JHAs and instructional SOPs are used to train employees about the hazards and controls for their jobs. Another method used by WE-R-SAFE is Failure Modes, Effects, and Criticality Analysis (FMECA) process to access hazards on new equipment and processes.

(See Attachment 6: Sample of a Job Hazard Analysis).

3. Hazard Analysis of Significant Changes

Explain how, prior to activity or use, you analyze significant changes to identify uncontrolled hazards and the actions needed to eliminate or control these hazards. Significant changes may include non-routine tasks and new processes, materials, equipment, and facilities.

New equipment designs and the selection of new equipment/materials are reviewed for their safety and health aspects by the engineering department with input by hourly associates and a review by the Safety and Health Manager. Larger capital projects are reviewed by the Environmental Engineer and the Safety and Health Manager. Once the change is completed new baseline analyses are routinely performed. We have a management of change process to review the hazards associated with new tools, equipment, chemicals, change in procedures, change in personnel, etc.

4. Self-Inspections

Describe your worksite safety and health routine general inspection procedures. Indicate who performs inspections, their training, and how you track any hazards through to elimination or control. For routine health inspections, summarize the testing and analysis procedures used and qualifications of personnel who conduct them. Include forms used for self-inspections.

Inspections are a central part of our safety and health program. WE-R-SAFE policy is to ensure that safety inspections, which identify unsafe conditions and unsafe behaviors, are conducted on a regular basis and non-conformance is eliminated as quickly and responsibly as possible. In operations and maintenance areas these inspections are performed weekly. Other areas, such as the warehouse are inspected monthly, and office areas quarterly. All non-conformances list a person responsible for corrective action and a completion date. Corrective action follow-up is ensured and communicated using a database program which is posted on the WE-R-SAFE intranet.

Special inspections examine specific hazardous conditions and are necessitated by such factors as regulatory requirement, hazard severity associated with the operation or piece of equipment (for example, safety showers, eye washes, fire extinguishers, LO/TO, confined space permits, machine safeguards, etc.).

Inspections that focus on observing employee behavior are also conducted in WE-R-SAFE operations. Each department uses a checklist of critical behaviors specific to their work environment. These checklists are used during the weekly observations to evaluate safe behaviors. Results of the weekly observations are tallied and communicated to managers, supervisors and employees in the area.

Those formally designated to perform inspections are provided with hazard recognition training, staring with the OSHA 10-Hour General Industry class and continuing with site specific hazard recognition training that includes the use of the forms and tracking process.

(R&D Safety Inspection Checklist, WE-R-SAFE Operations Safe Behavior Observations Report, Safety Inspection and Safe Behavior Observation SOP, Laboratory Safety Inspection Checklist, Safety Tracking example, Confined Space Entry Program, Control of Hazardous Energy Lockout/tagout Procedure, Electrical Safety Policy, Cranes and Hoists, WE-R-SAFE Powered Industrial Truck Licensing Program are available for review on site.)

5. Employee Reports of Hazards

Describe how employees notify management of uncontrolled safety or health hazards. Explain procedures for follow up and tracking corrections. An opportunity to use a written form to notify management about safety and health hazards must be part of your reporting system.

During their initial safety orientation, employees are introduced to the basics of hazard recognition and given copies of the Improvement Opportunity Form. They are instructed to use this form to alert Safety Teams, supervisors and SHO of near misses, safety and health hazards, ergonomics issues, and safety and health suggestions. The use of this form is encouraged during annual safety re-training in all departments. Once an opportunity form is received, SHO initiates an investigation of the alleged hazard. If a hazard exists, recommendations are made for hazard control. Every effort is made to involve the affected employees in feasible control solutions. The recommendations are then converted to action items with responsible parties and due dates which are tracked using our document control system. Employees are interviewed during safety inspections and asked about any hazards they are aware of and for input on improvement opportunities.

Another mechanism for employee hazard reporting is the Ergo Hotline. The WE-R-SAFE Ergo Hotline is available for employees to report ergonomic issues directly to the Safety and Health Office (SHO). The phone number (ERGO or 3746) is answered by the SHO Safety Administrative Coordinator, who in turn refers the caller's concerns to a SHO or Wellness professional for follow-up. The Ergo hotline is introduced during initial employee safety training and reviewed again at safety annual re-training.

We foster open communication and employees usually verbally report issues. WE-R-SAFE also has a direct hot line where associates may report issues, problems or concerns. This line is staffed by independent communications specialists 24 hours a day, 7 days a week, toll free, and is confidential.

(The Improvement Opportunity Form, ERGO Hotline Information are available for review on site.)

6. Accident and Incident Investigations

Describe your written procedures for investigation of accidents, near misses, first-aid cases, and other incidents. What training do investigators receive? How do you determine which accidents or incidents warrant investigation? Incidents should include first-aid and near-miss cases. Describe how results are used.

WE-R-SAFE maintains a formal written program in incident reporting and investigation. All accidents/incidents/near misses, occupational injuries and illness are required to be reported to supervision immediately. Every WE-R-SAFE employee and contractor receives training on how to report an incident in the New Employee Safety Training. Supervisors are formally trained in Accident Investigation.

All incidents are reported and documented using the Work-Related Incident Report Form, which is readily available on the WE-R-SAFE Intranet. Reporting is required for near misses and minor injury or illnesses as well to provide management and employees with the necessary tools to recognize trends, investigate and analyze the accident or illness, determine the cause, and establish corrective actions. The supervisor and employee must complete the Incident Report form within 24 hours of notification of the event, then fax the completed report to the Company Occupational Safety Specialist and the Company

Occupational Health Nurse. The supervisor must also notify the department manager to make him/her aware that an accident/injury has occurred.

In addition, SHO performs a follow-up on every report. The extent of the follow-up is determined by the nature of the incident. Follow-up may involve a JHA or an ergonomics risk assessment depending on the nature of the incident. Corrective actions identified as a result of any incident investigation are assigned a responsible party and due date and tracked in the system. Incidents and corrective actions are reviewed at department meetings. Incidents which occurred at other facilities, if applicable for WE-R-SAFE, are also reviewed to prevent similar occurrence at our facility.

Near-Miss reporting is done via the Opportunity Improvement Form. All employees are encouraged during their training to do near-miss reporting. The employee submits the form to the SHO where it is initially reviewed. The item is usually directed to the area Safety Team for corrective action and follow-up depending on the nature of the report. Corrective actions identified as a result of any improvement opportunity are assigned a responsible party and due date and tracked in the system.

(See attachment 6: Work-Related Incident Report Form. The accident/Incident Investigation Policy is will be available for review on site.)

7. Pattern Analysis

Describe the system you use for safety and health data analysis. Indicate how you collect and analyze data from all sources, including injuries, illnesses, near-misses, first-aid cases, work order forms, incident investigations, inspections, and self-audits. Describe how results are used.

The SHO monitors and reports statistical data on occupational injury and illness. The Safety and Health Manager analyzes trends and root causes. Historical data is compared on a monthly, quarterly and annual basis to identify and implement targets for corrective actions. The data is also given to the Site Safety Team to be posted. Trend analysis data and results are presented quarterly to top managers during the site Executive Committee Meetings, at the monthly safety representative meetings, and is posted on the Safety Departments Intranet web site. Site safety goals, objectives and action plans are presented every year in broad communications meetings. Continuous improvement corrective action is done based on statistics. An example of this is WE-R-SAFE's Ergonomics Program that included extensive training to all associates in 2001 and is continuing into 2002. A major capital appropriation was done 1998 to update equipment throughout the site to address ergonomics issues.

(Examples of statistical charts will be available for review on site.)

D. Hazard Prevention and Control

1. Engineering Controls

Describe and provide examples of engineering controls you have implemented that either eliminated or limited hazards by reducing their severity, their likelihood of occurrence, or both. Engineering controls include, for example, reduction in pressure or amount of hazardous material, substitution of less hazardous material, reduction of noise produced, fail-safe design, leak before burst, fault tolerance/redundancy, and ergonomic design changes.

Although not as reliable as true engineering controls, this category also includes protective safety devices such as guards, barriers, interlocks, grounding and bonding systems, and pressure relief valves to keep pressure within a safe limit.

Priority is given first to the elimination of any recognized hazards through engineering controls, which may include using equipment to eliminate manual operations or installing guards. The next step in this process is to develop, implement and enforce administrative programs. Personal Protective Equipment is used to supplement these procedures and to protect employees that cannot be better controlled with higher levels of interventions.

Our engineering controls include extensive use of noise dampening equipment, material handling equipment such as conveyor systems, and vacuum lifts.

2. Administrative Controls

Briefly describe the ways you limit daily exposure to hazards by adjusting work schedules or work tasks, for example, job rotation.

If engineering controls are not feasible to control a hazard, then we will next rely on various administrative controls. We have extensive work practice controls in place at the facility, and they are discussed in the next section. Examples of non-work practice administrative controls are those to limit exposure to repetitive motion disorders. We address this with a job rotation program combined with mandatory stretch beaks for machine operators as well as computer users. The computers are programmed to hibernate automatically for 5 minutes every hour and to present a stretch program for the associate.

3. Work Practice Controls

Describe and provide examples of your work practice controls. These include, for example, workplace rules, safe and healthful work practices, specific programs to address OSHA standards, and procedures for specific operations that require permits, labeling, and documentation. Identify major technical programs and regulations that pertain to your site, such as lockout/tagout, process safety management, hazard communication, machine guarding, and fall protection.

We have in effect numerous programs based on the requirements of the OSHA standards that address the hazards present at our facility. These programs include: Control of Hazardous Energy, Confined Space Entry, Fall Protection, Respiratory Protection, Hazard Communication for Hazardous Chemicals, Emergency Response, First Aid, Electrical, Powered Industrial Trucks, Hot Work, and Personal Protective Equipment. The individual written programs will be available for review by the OSHA team during the VPP onsite evaluation.

Our Safety and Health Manual contains not only the above programs but an extensive list of rules and work practices that employees receive training on and must comply with. Since we do not use any of the chemical listed in the OSHA Process Safety Management Standard (1910.119) above the minimum threshold quantities, we do not have a Process safety management program.

4. Personal Protective Equipment

Describe and provide examples of required personal protective equipment (PPE) your employees use and what PPE the OSHA team members will need to bring to your worksite.

Personal Protective Equipment (PPE) requirements are determined based upon task and hazard assessments or evaluations performed by the SHO, along with the employee and supervisor or area engineer. PPE may be available to employees for beyond compliance requirements. These Hazard Assessments are conducted in accordance with the OSHA standard 29CFR1910.132. Approved safety glasses, respirators, various types of gloves, safety shoes, hearing protection, face shields, goggles, aprons, fall protection equipment, etc. are examples of PPE recommended for specific jobs or tasks.

(Available for review on site are: Respiratory Protection Program, PPE Program, Certificate of Hazard Assessment, Examples of completed Certificates of Hazard Assessments, Noise and Hearing Conservation Program, Safety Glasses Authorization, and Safety Shoes Authorization.)

5. Enforcement of Safety and Health Rules

Describe the procedures you use for disciplinary action or reorientation of managers, supervisors, and other employees who break or disregard safety and health rules.

There is a progressive disciplinary system. The system allows steps to be skipped based on the severity of the violation. The disciplinary program addresses infractions by both management and hourly associates. Training is considered based on the specific situation and a determination as to its potential effectiveness in preventing recurrent infractions.

6. Preventive/Predictive Maintenance

Summarize your written system for monitoring and maintaining workplace equipment to predict and prevent equipment breakdowns that may cause hazards. Provide a brief summary of the type of equipment covered.

Numerous types of equipment such as manufacturing, material handling, laboratory, and facilities equipment are in operation at WE-R-SAFE. Preventive Maintenance (PM) programs are in place to support facility and factory equipment. These comply with ISO and QSR standards and are available upon request. Industrial hygiene equipment is maintained through regular service and calibration, according to manufacturer recommendations. Those instruments requiring annual service by an authorized representative are sent off-site for service.

An example of a PM program is within Manufacturing. The Maintenance Group is responsible for the repairs and preventive maintenance of this manufacturing equipment. Standard Operating Procedures, referred to as RTS's (Repetitive Task Systems) are used as instruction guides in the maintenance of this equipment. RTSs are used to perform this preventive maintenance. An RTS may be performed on a daily, weekly, monthly, or annual basis. The criticality and previous performance of a piece of equipment determines the frequency of the RTS. The Department Planner is responsible for the scheduling of these RTSs and to ensure their timely completion. Completed copies of these RTSs are filed in the Maintenance Shop. RTSs are controlled documents under specifications imposed by ISO 9002 Standards.

(Available for review on site are: Equipment list, Additional Detail of Equipment and Preventative Maintenance, Manufacturing Documentation Master Index, WE-R-SAFE Maintenance Control Document Master Index.)

7. Occupational Health Care Program

Describe your on-site and off site medical service and physician availability. Explain how you utilize the services of licensed occupational health care professionals. Indicate the coverage provided by employees trained in first aid, CPR, and other paramedical skills, their training, and available equipment.

Our corporate safety and health organization provides a full time onsite registered nurse, a wellness coordinator, one part-time contract registered nurse, an employee assistance program counselor, and two health and fitness specialists to service the health and wellness needs of our employees.

The Health Services Office is handicap accessible and open 5 days per week, Monday through Friday from 7:00 AM to 4:30 PM. Off-hours are covered by Occupational and Environmental Medicine (OEM) Clinic based at the local Hospital. Phone triage is available 24 hours per day for advice and disposition to appropriate medical treatment centers.

One day per week for two hours, we have an onsite physician from the OEM clinic. The physician is board certified in Occupational Medicine and the assigned physician of record for WE-R-SAFE.

The Health Services Office staff provides on-site medical response and first-aid, medical management of workers' compensation and medical disability cases, and

OSHA Log maintenance. The Health Services Office also manages the following site specific OSHA medical surveillance programs: Respiratory Protection Program Exams, Hearing Conservation Exams, Bloodborne Pathogens Program, Ergonomics Assessment Survey and Laser Surveillance Program.

The onsite occupational health nurse is a RN with 5 plus years of occupational experience and is a Certified Hearing Conservation Specialist under OSHA guidelines. She holds a certificate of completion from the Corporation Ergonomics Program. Her previous experience includes emergency room and critical care background. She is a certified first aid and CPR instructor for the National Safety Council.

The part time nurse is a RN, BSN with extensive occupational health background. In addition she has experience as a nursing school instructor, as well as emergency room and critical care experience. In addition she is PIC certified. She provides services to WE-R-SAFE two days per week.

A Wellness Coordinator provides administrative support to the Health Office. She has an AAS in Business and has extensive computer experience in data base information. She also has 20 years of service with WE-R-SAFE.

In addition to the above, the facility has a team of trained individuals who serve as the Medical Emergency Response Team (MERT). Three of the members are EMT's and one is a Paramedic. The balance of the group is trained in standard first aid, CPR and the use of our several automated external defibrillators (AED). Each member of the group carries a pager alerting them to any emergency that requires attention during each shift. The pagers are activated by the emergency response number (x777). The team then responds to the location and assists the employee until outside emergency personnel arrive or WE-R-SAFE nursing personnel arrives for disposition.

In addition to the above described services, the health and wellness group oversees a wide variety of on-site wellness programs including, but not limited to: Health Screenings, Lifestyle Modification/Education Programs, Fitness Center and Aerobics Classes. The health and wellness staff partners with the SHO staff in different aspects of our safety program such as ergonomic evaluations, hazard analysis of new processes, accident investigations, and recommending PPE.

8. Emergency Preparedness

Describe your emergency planning and preparedness system. Provide information on emergency drills and training, including evacuations.

Emergency Action Plans have been developed for each building. These plans include all aspects of emergency preparedness, including fire, medical, chemical, biohazards, confined space, bomb, and security. Evacuation drills are held at least annually on each shift.

Several groups that play an important role are the Emergency Evacuation Organization Teams, Medical Emergency Response Team, Occupational Health Resources, Security and Facilities. Evacuations are critiqued using the Emergency Drill Observer Checklist.

E. Safety & Health Training

Describe the formal and informal safety and health training programs provided for managers, supervisors and employees. Include supervisors' training schedules and information on hazard communication, personal protective equipment and handling of emergency situations. Describe how you verify the effectiveness of the training given. (Sample attendance lists and tracking methods, if any, also may be attached if desired.)

All new employees, including managers, receive a detailed safety and health orientation training program. This orientation addresses management's commitment to the safety and health of all workers and reviews in detail, the numerous safety and health rules, regulations, policies, procedures and especially expectations. A significant portion of time (2-3 hours) is devoted to each employee's "face-to-face" discussion with the Safety Manager to "make it personal," in that the employee's safety and health is paramount … you're accountable to yourself, your family and to each employee here. The essence of this initiative is for each employee to build personal relationships with each other and always challenge each other "to think before acting" on any aspect of the job. For those employees who make it through the initial 90 day training period, the safety and health training is enhanced through supervisory "on-the-job" safety and training and expectations of the department, along with the extensive training database found on the companies intranet.

Behavioral Safety Training is also utilized as a medium, but only as a "philosophy" and is seen as a good tool for everyone despite their time and tenure at this location. Line supervisors perform the work detail training and drive the "train the trainer" initiatives throughout this facility. On occasion, the company utilizes vendors and safety and health consultants to augment the established program.

Safety and health training is based on hazard assessments and both leading and lagging indicators. Associates are very involved in the development of the annual safety and health training process, along with providing meaningful input to the review of policies and procedures.

Most of our managers and supervisors attended a course that was customized to WE-R-SAFE and presented by an outside consultant. The course is called Basic Safety & Health Training for Managers & Supervisors. A copy of the letter from top management requesting attendance at the training and one of the handouts from the training are available upon request.

All safety and health training is tracked via the intranet system along with each employees personal training history. Where any defined process dictates demonstration of skills, the employee is observed by either the trainer and/or supervisor using established checklists. Furthermore, all employees are trained in hazard recognition and analysis techniques initially and annually thereafter.

Training effectiveness is verified through evaluations that include, but are not limited to, demonstration of skills, written tests, workshop activities, or other training exercises.

Although the SHO does most of the scheduling and tracking of safety training, it is ultimately the department manager's responsibility to ensure that all of his/her personnel have completed safety training required in their area (and/or specific to their job).

Management verifies their understanding of the extensive safety and health hazards at this site by being "very visible on the floor" during each shift. All managers' performance standards begin with safety and health as a critical element in their annual performance reviews.

F. Assurances

As part of this application, We-R-Safe is proud to submit these VPP assurances:

1. Compliance

We-R-Safe will comply with the *Occupational Safety and Health Act (OSH Act)* and correct in a timely manner all hazards discovered through self-inspections, employee notification, accident investigations, OSHA onsite reviews, process hazard reviews, annual evaluations, or any other means. We-R-Safe will provide effective interim protection, as necessary.

2. Correction of Deficiencies

Within 30 days, We-R-Safe will correct safety and health deficiencies related to compliance with OSHA requirements and identified during any OSHA onsite review.

3. Employee Support

We-R-Safe employees support the VPP application. At sites with employees organized into one or more collective bargaining units, the authorized representatives for each collective bargaining unit for the majority of the covered employees has either signed the application or submitted a signed statement indicating that the collective bargaining agent(s) support VPP participation. At non-union sites, management's assurance of employee support will be verified by the OSHA onsite review team during employee interviews.

4. VPP Elements

VPP elements are in place, and management commits to meeting and maintaining the requirements of the elements and the overall VPP.

5. Orientation

Employees, including newly hired employees and contract employees, will receive orientation on the VPP, including employee rights under VPP and under the *OSH Act.*

6. Non-Discrimination

We-R-Safe will protect employees given safety and health duties as part of We-R-Safe safety and health program from discriminatory actions resulting from their carrying out such duties, just as Section 11(c) of the *OSH Act* protect employees who exercise their rights.

7. Employee Access

Employees will have access to the results of self-inspections, accident investigations, and other safety and health data upon request. At unionized construction sites, this requirement may be met through employee representative access to these results.

8. Documentation

We-R-Safe will maintain our safety and health program information and make it available for OSHA review to determine initial and continued approval to the VPP. This information will include:

- o Any agreements between management and the collective bargaining agent(s) concerning safety and health.
- o All documentation enumerated under Section III.J.4. of the July 24, 2000 *Federal Register* Notice.
- o Any data necessary to evaluate the achievement of individual Merit or 1-Year Conditional Star goals.

9. Annual Submission

Each year by February 15, We-R-Safe will submit the following information to the designated OSHA Regional VPP Manager:

- o Participant Rates

  a. For the previous calendar year, the TCIR for injuries plus illnesses, and the DART.

  b. The total number of cases for each of the above two rates.

  c. Hours worked and estimated average employment for the past full calendar year.

- o Contractor Rates

As a construction industry applicant, We-R-Safe will submit data for all contractors performing work under our control in the DGA. The data will consist of:

a. The site's TCIR and DART for each contractor's employees.

b. The total number of cases from which these two rates were derived;

c. Hours worked and estimated average employment for the past full calendar year.

d. The appropriate SIC code for each applicable contractor's work at the site.

o Annual Evaluation

We will attach a copy of the most recent safety and health annual evaluation. It will include a description of any success stories, such as reductions in workers' compensation rates, increases in employee involvement, and improvements in employee morale.

10. Organizational Changes

Whenever significant organizational or ownership changes occur, We-R-Safe will provide OSHA within 60 days a new Statement of Commitment signed by both management and any authorized collective bargaining agents.

11. Collective Bargaining Changes

Whenever a change occurs in the authorized collective bargaining agent, We-R-Safe will provide OSHA within 60 days a new signed statement indicating that the new representative supports VPP participation.

____

Signed by Plant Manager.

Table 1. Site Employee Recordable Nonfatal Injury and Illness Case Incidence Rates

G. Rate Calculations and Tables

	A	B	C		E	F	G	H	I
Year	Total Work Hours	Total # Injuries	Total # Illnesses	Sum of Injuries and Illnesses	TCIR[2]	Total # DART Injury Cases	Total # DART Illness Cases	Sum of DART Injury and Illness Cases	DART[3] Rate
3 Years Ago (2006)	952,000	13	2	15	3.2	8	1	9	1.9
2 Years Ago (2007)	960,000	15	1	16	3.3	9	0	9	1.9
Last Year (2008)	930,000	11	0	11	2.4	5	0	5	1.1
3 Year Totals and Rates	2,842,000	39	3	42	3.0	22	1	23	1.6
2007 BLS Rates for NAICS 123456					4.3				3.1
Percent Below/Above BLS National Average[4]					31%				48%
Current Year to 12/12	410,000	5	0	5	2.4	2	0	2	1.0

Table 2. Site Applicable Contractor Recordable Nonfatal Injury and Illness Case Incidence Rates[5] (for work at your site only)

Name of Contractor									
NAICS Code for work at site									
	A	**B**	**C**	**D**	**E**	**F**	**G**	**H**	**I**
Year	**Total Work Hours**	**Total # Injuries**	**Total # Illnesses**	**Sum of Injuries and Illnesses**	**TCIR**	**Total # DART Injury Cases**	**Total # DART Illness Cases**	**Sum of DART Injury and Illness Cases**	**DART Rate**
3 Year s Ago (20XX)				0	0			0	0
2 Year s Ago (20XX)				0	0			0	0
Last Year (20XX)				0	0			0	0

Note: You do not have to submit applicable contractor rates with your application, but you must maintain them at the site for review by the OSHA VPP Team. Approved participants do submit applicable contractor rate data each year as part of their annual submission to OSHA.

ATTACHMENTS

1. Safety Policy and Vision

2. Organization Chart

3. Safety Roles and Responsibilities

4. WE-R-SAFE Performance Review Form

5. Previous Year's Annual Evaluation

6. Sample Job Hazard Analysis

7. Work-Related Incident Report Form

Attachment 1

Safety Policy and Vision.

As a core value, the safety, health, and well-being of our employees and clients is of paramount importance.

PRINCIPLE 1 – We conduct operations in a manner that protects the safety, health and well-being of our employees and our customers.

- Our top priority is the safety, health, and well-being of our employees and clients.
- State of the art safety and health programs and systems will be implemented to protect our employees. Mere compliance with applicable laws and regulations is not enough.
- Safety and occupational health will be integrated into all our work activities.
- Without fear of reprisal and discrimination, employees shall have the authority to stop work activities that would expose them to imminently dangerous hazards
- Emergency situations in the office and in the field will be handled appropriately.
- Our goal is to eliminate occupational injuries and occupational illnesses for our employees and protect them when elimination is impossible.

PRINCIPLE 2 – Everyone is committed to the process.

- Management will be accountable for safety and occupational health through demonstrated commitment to the safety and health management system.

- All employees will recognize that working safely is a condition of employment, and that they are accountable for their own safety and the safety of those around them.

- Resources will be available to achieve our goals.

- Potential risks in all of our operations (including inspections) will be addressed.

- All employees will be trained to perform their jobs safely.

- Off-the-job safety will be promoted to extend and reinforce safety and health consciousness.

PRINCIPLE 3 – Performance is measured and the results will be used toward on-going improvement of our Safety and Health Management System

- Effective performance measures will be utilized toward improving our safety and health management system.

- Behaviors, work processes, and management systems will be routinely examined.

- All incidents and near misses/hits will be investigated to determine contributing factors and to improve ongoing prevention efforts.

- Open communication regarding safety and health performance will be fostered at all levels.

Attachment 2

Organization Chart

Attach an organization chart for your facility.

Attachment 3

Safety Roles and Responsibilities

Plant Manager

- Develop and publish a Safety and Health Philosophy for the plant
- Develop and communicate Safety and Health Goals for the plant
- Implement all safety and health procedures and programs
- Communicate expectations and provide resources
- Evaluate and measure all management in meeting safety and health goals as stated in the various policies referenced in the application
- Conduct facility inspections
- Reviews all incident reports

Operations Manager

- Actively demonstrates support for the Safety and Health Philosophy and Goals
- Schedules training
- Schedules annual physicals
- Tracks training and checks compliance
- Coordinates safety meetings
- Participates on the Safety / VPP Committee
- Completes quarterly LOTO audits
- Assists in all incident investigations

Maintenance Manager

- Actively demonstrates support for the Safety and Health Philosophy and Goals
- Oversees the contractor safety program
- Implements the facility's preventive maintenance program
- Coordinates powered industrial truck and overhead crane inspections
- Assists in all incident investigations

Safety and Health Manager

- Actively demonstrates support for the Safety and Health Philosophy and Goals
- Updates all procedures and programs as needed
- Assures compliance with all safety and health procedures and programs
- Maintains Emergency Response Plan
- Interpret safety and health compliance regulations and policies
- Resolve identified heath and safety issues
- Reviews safety and health trends using both leading and lagging indicators
- Assists in all incident investigations

Safety / VPP Committee

- Meet on a regular basis (Monthly)
- Work with management, and the Safety and Health Manager to resolve safety issues
- Actively seek out employee safety and health issues to address
- Develop and communicate safety and health goals for the facility

Employees

- Follow all safety rules and instructions (wear proper PPE, follow JHA's, etc.)
- Work in a safe manner
- Support the safety team and actively participate in the safety program
- Participate in accident and near miss investigations
- Report unsafe acts/conditions
- Maintain housekeeping in their immediate work areas

Attachment 4

WE-R-SAFE Performance Review Form

NAME	EVALUATION PERIOD
TITLE	DEPARTMENT UNIT

Note: Retain each update in support of annual summary.

RATING SCALE

(This scale should be used throughout the document, in those areas designed for an evaluation of performance.)

Rating	Rating Description	Definition of Rating
0	Unacceptable Performance	Performance does not meet the expectations of the position and/or performance objectives for the time in the position. Use of appropriate progressive corrective action is required. (Requires comments and specific examples of performance in action plan).
1	Performance Below Expectations	Performance does not consistently meet the expectations of the position and/or performance objectives for the time in the position. Use of appropriate performance enhancement plan is required. Opportunity for improvement. (Requires comments and specific examples of performance in action plan).

2	Performance Somewhat Within Expectations	Performance somewhat meets the expectations of the position and/or performance objectives for the time in the position. Use of appropriate competencies requires development. Opportunity for improvement. (Requires comments and specific examples of performance in action plan).
3	Performance Within Expectations	Performance meets expectations of the position and performance objectives for the time in the position. Demonstrates a skillful use of most competencies to achieve results.
4	Performance Above Expectations	Performance consistently meets and occasionally exceeds expectations of the position and performance objectives for the time in the position. Demonstrates a skillful use of competencies to achieve results. (Requires comments and specific examples of performance in action plan).
5	Excellent Performance	Performance frequently exceeds expectations of the position. Demonstrates the use of the necessary competencies to achieve results. Frequently demonstrates a higher level of performance than required of position. (Requires comments and specific examples of performance in action plan).
6	Exceptional Performance	Performance always exceeds expectations of the position. Consistently demonstrates the use of the necessary competencies to achieve exceptional results. Demonstrates a consistently higher level of performance than required of position. (Requires comments and specific examples of performance in action plan).
7	Too New to Rate	Person has been in position for less than three months. Insufficient observations to be able to rate performance.

Evaluation criteria

#	Item	Description	Mid-Year Rating (0-7)	Yearly Rating (0-7)
1	EHS	Works consistently with the principles and elements of the EHS Policy. Demonstrates competence and commitment to safe and environmentally sound work habits. Shows that actively cares about safety and environmental aspects in the work area and of other colleagues. Follows all applicable safety and environmental rules. Actively participates and gets involved in safety and environmental related initiatives and activities.		
2	Quality	Strives to provide customers with the highest quality product. Demonstrates an understanding of quality requirements in applicable processes, equipment and procedures. Produces accurate and thorough results. Identifies and corrects quality problems.		
3	Teamwork	Uses good verbal and written communication skills. Actively participates in team activities. Treats people with respect and dignity. Handles conflict resolution in a constructive manner. Exhibits a willingness to adapt to changing work priorities. Exhibits a willingness and ability to rotate through different jobs.		

4	Integrity	Strive to complete all job requirements with the highest degree of effectiveness. Daily conduct reflects high standards both professionally and ethically. Behaves openly and honestly with colleagues and customers. (3 ratings choices in this dimension- 0, 3 or 6)		
5	Innovation	Demonstrates a willingness to become involved, initiates positive actions and ideas. Pursues challenges in work environment with enthusiasm and creativity.		
6	Customer Focus	Focuses on customer needs and satisfaction. Constantly strives to produce superior products and service. Anticipates needs or responds quickly to changing conditions.		
7	Respect for People	Treats all others with respect and dignity. Responds positively to coaching and constructive feedback. Assists other work areas or team members without being asked.		
8	Leadership / Community	Steps forward in achieving difficult goals. Shows an understanding of what needs to happen to achieve goals. Recognizes and evaluates problems, performs the appropriate investigation to reach sound conclusions.		
9	Performance	Performs all duties with the highest degree of effectiveness and in a cost-conscious manner. Works to improve efficiency of operations. Employee links behavior to facility and team goals. Makes full and effective use of time.		
		Current Period Total:		
Note: Average of ratings		Current Period Average:		

Performance Dimensions Action Plan

This space can be used to highlight accomplishments, major strong points and areas for improvement for this review period. You must indicate the performance dimension and the reasons for the rating given for all ratings except "Performance within Expectations" and "Too New To Rate".

This space can be used to highlight areas of development for the next review period (i.e., courses, seminars, projects assignments geared towards enhancing the colleague's skills and knowledge).

Development Plan:

SIGNATURES

By signing below, we acknowledge that the colleague has reviewed and received a copy of the performance evaluation. It does not necessarily signify acceptance or agreement with the evaluation. If the colleague has written comments regarding the evaluation, a separate sheet should be attached.

Colleague: ____________________ Date: ____________

Leader: ____________________ Date: ____________

Attachment 5

Previous Year's Annual Evaluation

We-R-Safe

SAFETY AND HEALTH MANAGEMENT PROGRAM REVIEW

REPORT FOR YEAR

2008

PREPARED BY: Norman Lynch, Safety and Health Manager

APPROVED BY: Brian Smith, Plant Manager

2008 SAFETY AND HEALTH PROGRAM MANAGEMENT REVIEW

INDEX

PROGRAM ELEMENT

1.0 Updated Company Information

2.0 Status of Safety and Health Corrective Action Reports

3.0 Injury/Illness Rates

4.0 Management Leadership and Employee Involvement

4.1 Management Commitment

4.2 VPP Commitment

4.3 Planning

4.4 Written Safety and Health Management System

4.5 Top Management Leadership

4.6 Authority and Resources

4.7 Line Accountability

4.8 Contract Workers

4.9 Employee Involvement

4.10 Safety and Health Management System Evaluation

5.0 Worksite Analysis

5.1 Management Understanding

5.2 Industrial Hygiene

5.3 Pre-Use Analysis

5.4 Hazard Analysis

5.5 Routine Inspections

5.6 Employee Hazard Reporting Program

5.7 Accident/Incident Investigations

5.8 Trend Analysis

6.0 Hazard Prevention and Control

6.1 Certified Professional Resources

6.2 Hazards Elimination or Control

6.3 Process Safety Management

6.4 Preventive Maintenance

6.5 Hazard Correction Tracking

6.6 Occupational Healthcare Program

6.7 Disciplinary Program

6.8 Emergency Procedures

7.0 Safety and Health Training

7.1 Supervisors

7.2 Employees

7.3 Emergency Response

7.4 PPE

7.5 Managers

8.0 Significant Changes in Safety & Health Management System

9.0 Success Stories

10.0 Summary of Recommendations

1.0 UPDATED COMPANY INFORMATION

Company Name: We-R-Safe Company

Site Address: Anytown, USA

Plant Manager: Brian Smith

Site VPP Contact: Norman Lynch

Telephone Number: (555) 123-6789

FAX Number: (555) 467-8378

E-mail Address: norman@wersafe.com

Union Information: Non-Union

2.0 STATUS OF SAFETY AND HEALTH CORRECTIVE ACTION REPORTS

Following is a status summary of previous Safety and Health Corrective Action Requests:

	2008	2007	2006	TOTAL
Total SHCARs	18	15	9	42
#Complete	14	15	9	38
#In-Progress	4	0	0	4

Progress on recommendations from previous Safety and Health Management Program Reviews continues to be very good.

3.0 INJURY/ILLNESS RATES

	A	B	C	D	E	F	G	H	I
Year	Total Work Hours	Total # Injuries	Total # Illnesses	Sum of Injuries and Illnesses	TCIR[2]	Total # DART Injury Cases	Total # DART Illness Cases	Sum of DART Injury and Illness Cases	DART[3] Rate
3 Years Ago (2006)	952,000	13	2	15	3.2	8	1	9	1.9
2 Years Ago (2007)	960,000	15	1	16	3.3	9	0	9	1.9
Last Year (2008)	930,000	11	0	11	2.4	5	0	5	1.1
3 Year Totals and Rates	2,842,000	39	3	42	3.0	22	1	23	1.6
2007 BLS Rates for NAICS 123456					4.3				3.1
Percent Below/Above BLS National Average[4]					31%				48%
Current Year to 12/12	410,000	5	0	5	2.4	2	0	2	1.0

EFFECTIVENESS OF CURRENT PROGRAM

Although there was a slight increase in the number of recordable injuries from 2005 to 2006, the number and injury rates have dropped significantly from 2006 to 2007. This is attributed to the focus on safety and health brought about by our commitment to becoming a VPP Star worksite.

A review of the OSHA 300 Log and Incident Investigation Forms, which are maintained by the Safety and Health Manager, indicates that all reported injuries were classified and recorded properly.

RECOMMENDATIONS FOR IMPROVEMENT

None

4.0 MANAGEMENT LEADERSHIP AND EMPLOYEE INVOLVEMENT

4.1 MANAGEMENT COMMITMENT

ACTUAL PERFORMANCE/CURRENT YEAR ACTIVITIES

Management commitment to safety is spelled out in the plant's Safety Principles, which were updated in 2002 to incorporate WE-R-SAFE Company's Safety and Health Policy. Management also establishes annual S&H goals and objectives (discussed in Sections 4.3 and 4.5).

Plant management's commitment to safety continues to be demonstrated through the many man-hours and resources committed to such activities as the Plant Safety Committee (PSC), and Area Safety Teams. Management also shows its commitment to safety by empowering department operators to shut down operations, without fear of reprisal, if in their judgment the process is judged to be unsafe.

Management commitment was also demonstrated in 2008 by continued support of the behavior-based safety process at the site. The We-Share Team is currently comprised of members representing operating and maintenance areas of the plant. The team is working to reduce injuries by identifying barriers to safe behavior. Management has strongly supported this process, allowing time for training of observers and observation of tasks with the purpose of identifying and addressing at-risk behaviors and providing positive feedback for safe behavior. See Section 5.4 for additional information.

Management provides all necessary PPE, tools, and equipment to safely perform work.

EFFECTIVENESS OF CURRENT PROGRAM

Management commitment continues to be adequately demonstrated through provision of personnel and funds to ensure an effective safety and health program.

RECOMMENDATIONS FOR IMPROVEMENT

None.

4.2 VPP COMMITMENT

ACTUAL PERFORMANCE/CURRENT YEAR ACTIVITIES

The plant continued to demonstrate commitment to VPP through the following activities:

- Safety and Health Manager participated on two Regional VPPPA conference workshops and also attended two local STAR presentation ceremonies
- Safety and Health Manager attended the National VPPPA Conference in 2008
- We had a suggestion program to arrive at a new VPP slogan. The winning choice was: We-R-Stars

The site has also informally assisted and offered to mentor local firms who wish to pursue VPP participation using our recent experiences in that regard as an example.

EFFECTIVENESS OF CURRENT PROGRAM

Commitment to the principles and goals of the Voluntary Protection Program continues.

RECOMMENDATIONS FOR IMPROVEMENT

None.

4.3 PLANNING

ACTUAL PERFORMANCE/CURRENT YEAR ACTIVITIES

S&H planning elements include the following:

- Total Safety Key Emphasis Areas
- Salaried Employee Performance Planning and Appraisal Process
- Capital Plan
- S&H Corrective Action Request (SHCAR) process
- Compliance Planning

Total Safety Key Emphasis Areas are established and communicated at the beginning of each year by the S&H Manager. 2008 Key Emphasis Areas included:

- Behavior-based safety activities, implementation of Management Hey Safety
- Elements, reduction of At-Risk Behaviors related to fall protection and use of PPE
- Increased use of Job Safety Analysis for non-routine tasks
- Ergonomics Program implementation
- Reinforce compliance with fundamental safety procedures
- Improve followup to S&H-related findings and recommendations
- Improve contractor safety performance

Planning for Technical Services is largely based on the Capital Plan which is used to captured S&H-related projects for prioritization based on risk, implementation scheduling, and available capital and resourcing.

The plant's S&H Corrective Action Request tracking system is used to capture recommendations which result from incidents and audits; these flow into the Capital Plan and other planning mechanisms as appropriate.

All S&H requirements affecting WE-R-SAFE have been identified and an owner established for each. A mechanism for implementation planning for new S&H requirements is also in place as part of the plant's S&H Compliance Process.

EFFECTIVENESS OF CURRENT PROGRAM

Total Safety Key Emphasis Areas have been very effective in providing improvement objectives for the safety and health program. Significant progress was made in 6 of 7 Key Emphasis Areas in 2008. Key Emphasis Areas are carried over into the following year as appropriate.

The WE-R-SAFE Salaried Employee Performance Planning and Appraisal Process, was implemented in 2008; it is too early to assess effectiveness in setting annual goals and objectives for management and technical personnel.

The Capital Plan has been effective as the primary planning tool for the Technical Services Department (TSD) effort. S&H issues have typically been a high priority, and a significant portion of TSD effort has been devoted to S&H-related projects. Effectiveness has been improved through continuous planning vs. annual planning, which was instituted as part of the Capital Project Management process. Effectiveness of WE-R-SAFE's capital project planning/management process is yet to be determined. Capital work was significantly curtailed in 2008 due to business and economic conditions.

SHCAR Tracking has been effective in capturing and funneling S&H recommendations into work plans and other planning mechanisms. SHCAR progress tracking is easily monitored via the on-line SHCAR system. A record 14 SHCARs were completed in 2008 from incident investigations, employee concerns, and audits.

The Compliance Planning process was used successfully to implement significant changes to the plant's contractor safety program and OSHA Recordkeeping requirements.

RECOMMENDATIONS FOR IMPROVEMENT

None.

4.4 WRITTEN SAFETY AND HEALTH MANAGEMENT SYSTEM

ACTUAL PERFORMANCE/CURRENT YEAR ACTIVITIES

The plant's written Safety and Health Management System is contained in (4) manuals which together include over 50 S&H procedures covering almost all applicable requirements:

*S&H Systems Procedures

*Safety Procedures

*Health Procedures

*Mechanical Integrity Procedures

In accordance with the plant's S&H Compliance Process, each S&H requirement is assigned to a Requirement Owner who is responsible for developing and updating applicable procedure(s). Written procedures are used both for initial and refresher training; they also include an audit checklist for performing internal audits to verify that plant practice is in accordance with applicable procedures.

In 2008, a significant number of S&H procedures were updated and reissued:

- S&H Procedure Management
- S&H Corrective Action Tracking
- Compliance Audit Program
- Permit-Required Confined Space Entry
- Fire Permits
- Lockout/Tagout of Electrical/Mechanical Forms of Energy

- Ladders, Scaffolds, and Powered Aerial Platforms
- Incident/Injury Investigations and Reporting
- Fall Protection
- Contractor Safety Procedure
- Contractor Approval and Selection (new procedure)
- Job Safety Analysis
- Emergency Response Plan

The frequency for procedure review and update was extended from 3 to 5 years (more often if significant changes are made). An annual plan for procedure update is developed. Procedures are also revised to reflect changed requirements, incident investigation recommendations, audits findings, or other input. Procedures are reviewed by affected parties (including operators and mechanics) for input prior to issue.

Plant Safety Rules are contained in plant S&H procedures and the Guidelines for Contractors handbooks. Department-specific safety and health rules for operating units are included in their Standard Operating Procedures.

Basic safety and security rules are included in new employee safety orientation and training, as well as in orientations for contractors and visitors/drivers/service personnel.

In 2008, an electronic on-line database for S&H procedures was established on the plant Intranet to facilitate access from anywhere in the plant. Thirteen controlled paper copies are also maintained.

EFFECTIVENESS OF CURRENT PROGRAM

Plant S&H procedures adequately cover virtually all applicable requirements. They are generally well-written and thorough. Audits of each procedure are performed periodically to verify compliance and identify improvement opportunities. There were several that were found to not have been reviewed for over 6 years.

RECOMMENDATIONS FOR IMPROVEMENT

1. Review and update those S&H procedures that have not been reviewed withing the past five years.

ISSUED TO: J. King

Priority: H

EDC: 12/31/2009

4.5 TOP MANAGEMENT LEADERSHIP

ACTUAL PERFORMANCE/CURRENT YEAR ACTIVITIES

Top management at the site demonstrates safety leadership in the following ways:

- Establishes plantwide safety goals and objectives, both leading and trailing. In 2008, the following goals were established as part of the plant's Incentive Program:
    - Completion of (400) behavior-based safety observations
    - Completion of (24) S&H procedure audits
    - Completion of 100% of S&H training
    - Completion of 80% of corrective actions from safety observations

  All of these goals were met.

- Participation in the Administrative Safety Committee (ASC), comprised of the Plant Manager and staff, Safety/IH Supervisor, the Chair- and the Vice-chairperson of the PSC, and the We-R-Safe Team Leader. The ASC meets at least monthly to monitor progress against Total Safety Key Emphasis Areas and goals, facilitate plant-wide safety efforts, and help address issues raised by the PSC, as well as to review and improve incident investigation reports. Minutes are distributed plantwide via e-mail. A charter is reviewed and updated annually.

- Participation in the monthly Safety & Housekeeping Inspection Program; in 2008, the plant manager asked that all supervisors also participate in these inspections as well as in conducting audits of S&H procedures.
- Monthly Plant Informational meetings, held for all employees, begin with a review of S&H performance (both leading and trailing indicators) against goals.
- Open-door policy and clear lines of communication between employees and management personnel.

Management remains committed to providing the leadership to drive continuous improvement in our plant's We-R-Safe process through facilitation of employee involvement in order to achieve the plant's vision of an incident-free work environment.

EFFECTIVENESS OF CURRENT PROGRAM

The overall effectiveness of top management leadership in safety continues to be good. This is also substantiated by the plant's ability to demonstrate continued good safety and environmental performance.

The ASC is effective in setting S&H goals, monitoring progress, and removing barriers to help achieve them.

RECOMMENDATIONS FOR IMPROVEMENT

None.

4.6 AUTHORITY AND RESOURCES

ACTUAL PERFORMANCE/CURRENT YEAR ACTIVITIES

All employees have the authority to stop or refuse to perform any job, including starting up or operating units, without fear of reprisal, which in their judgment involves an unacceptable risk, until appropriate corrective/preventive actions have been taken. Supervision generally has adequate authority to correct unsafe conditions and practices in their area, although WE-R-SAFE top management approval is required for any capital expenditures.

All plant personnel are considered as resources for the facility Safety and Health program because each individual plays an important role and are afforded the opportunity to actively participate in Safety and Health activities outside the scope of their routine job.

Plant resources that are dedicated largely to Safety and Health are:

- Safety and Health Manager
- Safety Engineer
- Industrial Hygienist
- Contractor Supervisor- reports to Operations Manager
- We-R-Safe (behavior-based safety process) Team Leader (also serves as IH technician reporting to Industrial Hygienist)
- Plant and Department Training Coordinators (full-time for manufacturing, part-time for others)

The Safety and Health Manager reports directly to the Plant Manager.

In addition to a Plantwide Training Coordinator, each department has a Department Training Coordinator.

Financial resources are allocated to support existing Safety and Health programs. This is planned for during the annual budgeting process. Capital for safety and health projects is allocated as part of the Capital Planning Process. In 2008, approximately $68k was approved for safety and health-related capital projects vs. $270k in 2007, down largely due to business and economic conditions.

EFFECTIVENESS OF CURRENT PROGRAM

The level of safety authority provided to plant personnel is considered to be adequate. Resources dedicated to the safety and health program, although reduced from earlier years, remain adequate.

RECOMMENDATIONS FOR IMPROVEMENT

None.

4.7 LINE ACCOUNTABILITY

ACTUAL PERFORMANCE/CURRENT YEAR ACTIVITIES

Mechanisms for establishing line management accountability for safety & health activities include:

- S&H Safety Policy and Programs (primary responsibilities)
- S&H procedures
- S&H Corrective Action Request (SHCAR) process

All plant management and supervision are expected to spend a significant portion of their time on safety. Primary responsibilities of management and supervision are included in S&H Safety Policy and Programs. Responsibilities of all affected parties are established as part of each S&H procedure.

Accountability for follow-up of S&H improvement/corrective actions from audits and incidents is established through the S&H Corrective Action Request (SHCAR) process.

In 2008, WE-R-SAFE's Salaried Employee Performance Planning and Appraisal Process, which includes a section on S&H, was instituted.

EFFECTIVENESS OF CURRENT PROGRAM

Mechanisms for establishing and enforcing line management accountability for safety & health activities are deemed adequate at this time.

RECOMMENDATIONS FOR IMPROVEMENT

None.

4.8 CONTRACT WORKERS

ACTUAL PERFORMANCE/CURRENT YEAR ACTIVITIES

Contractor Recordable incidents in 2008 were as follows:

- Eye irritation- mechanical contractor sprayed with sludge
- Facial laceration- scaffold erector struck by board
- Loss of consciousness – heat stress

Contractor injury/illness rates have increased in the past two years after almost 1000 days without a Recordable incident. This has been attributed to a large amount of turnover in regular contractor personnel. Total injuries (including minor) were down, due in part to a reduction in the number of contractors working in the plant, however the number of Recordable incidents remains unacceptably high. Safety observations continue to focus on contractors in an effort to improve this performance, with (74) observations (18% of total) conducted on contractors in 2008, which is equivalent to the relative number of hours worked by contractors and employees.

The Contractor Safety Program is covered in two procedures:

- Contractor Approval and Selection
- Contractor Safety Procedure

The Contractor Approval and Selection Procedure covers initial review and approval of contractors working in operating areas along with periodic re-evaluation to verify injury/illness and Experience Modification (EMR) rates and also that safety and health and training programs are in-place.

The Contractor Safety Procedure includes the following elements:

- Safety Orientation
- Work Permit
- Monthly inspections of contractor work areas
- Weekly construction site inspections
- Training for those who supervise contractors to ensure proper oversight

This process focuses first on bringing in the safest contractors, then to properly supervise and periodically review performance.

The plant participates in a local consortium of companies focused on improving contractor safety at all sites. This consortium sponsors a (4)-hour initial/annual refresher safety awareness course (with test requiring 80% to pass) which is required for all contractors who work in member companies' facilities.

Contractors are not permitted to perform physical work unless they have received the site safety orientation which includes viewing the safety orientation program

(CBT or video), reviewing the plant contractor safety booklet, and undergoing testing to display written verification of understanding. Annual reorientation is required. Consequences of non-compliance with plant safety rules, including dismissal, are covered in the plant orientation.

Monthly contractor safety meetings are conducted for regular contractors covering the following topics:

- Contractor safety orientation refresher
- Fall protection, ladders & scaffolds
- Electrical safety, lockout/tagout
- Fire permits & fire extinguishers
- Confined space entry, excavations
- Hazard communication, industrial hygiene, & respiratory protection
- PPE
- Process unit overviews & tours (including chemical hazards)

Monthly meetings also allow for discussion of safety issues of concern.

All plant visitors receive a plant safety orientation provided by the Guard. A 6-question quiz is also administered to verify understanding and English proficiency.

Contractor safety was an emphasis area in 2008, and the following activities were completed:

- Audit of Contractor Safety Procedure / Work Permit
- Contractor Safety Procedure / Work Permit reviewed and updated

- New procedure for Contractor Approval and Selection developed
- New Approved Contractor on-line database established
- Training for Contractor Management Representatives (CMRs) who supervise outside contractors
- Audits conducted of primary contractors (>1000 hrs/year) to verify safety and health program and training.

Since completion of these activities in the first half of the year, there were no contractor incidents for the remainder of the year, and overall contractor safety appears to be improving.

EFFECTIVENESS OF CURRENT PROGRAM

Overall, the contractor safety program contains all necessary elements, and behavioral safety observations indicate a high level of safety awareness and compliance among contractors. While the total number of contractor injuries (including minor injuries) is low (4 reported in 2008), three were classified as Recordable. Continued emphasis on contractor safety is needed in 2009 to maintain the improved trend experienced in the second half of 2008.

RECOMMENDATIONS FOR IMPROVEMENT

1. Include audit of contractor safety procedures in S&H Audit Schedule.

ISSUED TO: J. King

PRIORITY: M

EDC: 12/31/2009

4.9 EMPLOYEE INVOLVEMENT

ACTUAL PERFORMANCE/CURRENT YEAR ACTIVITIES

WE-R-SAFE has a strong tradition of employee involvement in safety and health. In addition to safety committees, formal mechanisms for employee involvement in safety and health programs include:

- S&H Procedure Reviews
- Project Safety Reviews
- Behavior-based safety process
- Job Safety Analyses
- Safety & Housekeeping Inspections
- Hazard Correction Requests
- Incident Investigation teams
- Plant Emergency Brigade
- Process safety management activities
- Plant Training Team

See appropriate section(s) for descriptions of these programs.

Several different safety and health committees are active at the WE-R-SAFE, including:

- Plant Safety Committee (PSC)
- Administrative Safety Committee (ASC)
- We-R-Safe Steering Team

The PSC serves as a focal point for employee involvement. The PSC is comprised of a Chairperson, Vice-chairperson, a representative from each department, We-R-Safe facilitator, a supervisor representative and a Safety Department representative. Any employee may be nominated to be Vice-chairperson (selection made by the ASC), who then becomes the chairperson the following year. Membership is voluntary with terms lasting one year or more depending on department selection. The PSC meets for 2–4 hours monthly, and addresses and communicates plant-wide safety issues, including status review of Hazard Correction Requests. They also select a Safety Leader of the Month. The PSC helps address safety issues that go beyond any one department or have no clear owner. Agendas and minutes are distributed plant-wide via e-mail. Subgroups are sometimes used to work issues. Charter review and update and training on key topics (such as Safety and Housekeeping Inspections) are performed annually.

The WE-R-SAFE. Steering Team is comprised of a part-time facilitator and broad plant representation, primarily hourly. The WE-R-SAFE. team is charged with managing the behavior-based safety process. They meet periodically to evaluate safety observation data and develop action plans to address barriers to safe behavior. They communicate to the plant regularly via minutes, newsletters, and participation in safety meetings. Status reports are provided routinely to the PSC.

EFFECTIVENESS OF CURRENT PROGRAM

Overall effectiveness of employee involvement in safety and health programs continues to be high. A sizeable number of safety concerns are identified and addressed throughout the year through employee participation. Department safety meetings, and other sources of training and communication have encouraged employees to participate in safety awareness and improvement efforts.

Effectiveness of employee involvement in the WE-R-SAFE. process has been particularly high and continues to identify barriers to safe behavior by both employees and contractors. A reduction in the number of observations in the second half of 2008 is cause for concern, and appears to be related to overall concerns regarding the business climate which declined during this time, as well as the departure of several active observers.

Concerns regarding effectiveness of the Plant Safety Committee were addressed in 2008 through development of an action plan which improved the visibility and communication of the PSC, including monthly visits to areas of the plant to assess safety and health perceptions, identify concerns, and communicate PSC and other plant safety efforts. Based on employee interviews, this has increased perceived PSC effectiveness. Attendance at monthly PSC meetings was 80% for the year, a new safety process measure being instituted in 2008.

The activity level and effectiveness of Area Safety Teams varies from department to department and additional supervisory emphasis is needed to ensure effectiveness in all areas.

The WE-R-SAFE. team has been extremely effective in implementation of the behavior based safety process and identifying and helping to address barriers to safe behavior, however participation dropped in the latter part of 2008.

RECOMMENDATIONS FOR IMPROVEMENT

None

4.10 SAFETY AND HEALTH MANAGEMENT SYSTEM EVALUATION

ACTUAL PERFORMANCE/CURRENT YEAR ACTIVITIES

The annual Safety and Health Management Program Review was conducted by a team comprised of S&H personnel with input from affected plant employees. Each program element was evaluated based on new VPP requirements. Evaluation and improvement planning efforts are coordinated with those of the Plant Safety Committee and Administrative Safety Committee to ensure consistency. The ASC reviews the final report prior to issue.

S&H Corrective Action Requests are generated for recommendations from each annual Program Review and are tracked per S&H Corrective Action Tracking.

EFFECTIVENESS OF CURRENT PROGRAM

The current process for conducting the annual Program Review is deemed to be very effective in evaluating the safety and health management system and identifying opportunities for improvement. Progress on addressing improvement recommendations from past reviews continues to be high.

RECOMMENDATIONS FOR IMPROVEMENT

None.

5.0 WORKSITE ANALYSIS

5.1 MANAGEMENT UNDERSTANDING

ACTUAL PERFORMANCE/CURRENT YEAR ACTIVITIES

Management understanding of safety and health conditions at the site is established in the following ways:

- Comprehensive surveys
- Industrial Hygiene surveys
- Pre-Use Analysis
- Hazard Analysis
- Routine Inspections
- Employee Hazard Reporting Program
- Accident/Incident Investigations
- Trend Analysis

See appropriate sections for a review of these elements.

Comprehensive surveys include the following:

- S&H procedure self—audits
- Process safety/risk management program audits—annual
- WE-R-SAFE Safety and Health—annual
- Insurance Company Assessments

The following comprehensive surveys/audits were conducted in 2008:

- Risk Management Program—Done by Consultant
- Audit of Risk Management Program—annual
- Plant wide equipment guarding survey
- Plant wide electrical equipment safety inspection
- Plant wide safety shower/eyebath survey

The S&H Audit program was fully re-instituted in 2008. Twenty-three of twenty-four audits scheduled were completed. There were more audits done in 2008 than any other year. Audits completed were:

- Lockout/Tagout—quarterly
- Confined Space Entry
- Fall Protection
- Management of Change
- Fire Permit
- Respiratory Protection
- Maintenance of Slings, Hoists & Lifting Devices

- Contractor Safety
- Ladders, Scaffolds & Powered Aerial Platforms
- Fire Protection Impairment Control
- Hearing Conservation
- Chemical Use Approval

These are in addition to the process safety management self-audits listed above. A schedule of internal audits/surveys is maintained by the Safety and Health Manager. In 2008, a quarterly S&H audit closing meeting was instituted to review results of all completed audits, including findings and assignment of responsibility to address.

All recommendations from comprehensive surveys are tracked in the SHCAR Database in accordance with the S&H Corrective Action Tracking, and are fully accessible to all plant personnel on the on-line SHCAR tracking and reporting database. In 2009, (173) SHCARS from comprehensive surveys were completed, (172) new recommendations were added and (203) recommendations remained open at year-end.

EFFECTIVENESS OF CURRENT PROGRAM

Effectiveness of the S&H Audit Process, particularly process safety management program self-audits, are generally effective in verifying compliance with internal and external requirements. Emphasis on self-audits will continue into 2009.

The WE-R-SAFE insurance company assessment was conducted in May, 2008 and focused primarily on fire protection issues.

RECOMMENDATIONS FOR IMPROVEMENT

None.

5.2 INDUSTRIAL HYGIENE

ACTUAL PERFORMANCE/CURRENT YEAR ACTIVITIES

Recognition—An inventory of hazards, chemical and physical, has been established for each operating unit which are identified in the Status Report & Assessment Strategy (SR/AS). The most recent SR/AS was conducted in 2007.

Evaluation—Ongoing IH monitoring samples are taken to assess exposure to chemicals and noise. Sample collection is performed by the IH technician under supervision of the plant Certified Industrial Hygienist. Most IH samples are collected and analyzed by recognized methods; in cases where no recognized method exists, validated in-house methods are used. In all cases sample analysis is performed by an AIHA—certified lab. Very few monitoring results indicate exposure over established occupational exposure limits/guidelines. Respiratory protection is employed in all cases where limits/guidelines are exceeded and in cases where there is anticipated exposure. The Status Report and Assessment Strategy (SR&AS) report is developed as a summary of exposure monitoring results as well as strategy for future monitoring.

Control—A number of written Industrial Hygiene programs have been developed to facilitate compliance with OSHA mandated programs. In addition, there are established control methods in use to reduce exposure and exposure potential. These include engineering controls, work practices and personal protective equipment.

Department materials lists and statistical evaluation of exposure data are used for exposure assessments and for establishing a sampling strategy.

EFFECTIVENESS OF CURRENT PROGRAM

The plant I.H. program provides an effective combination of hazard assessment through exposure monitoring with data analysis and professional judgement. Programs are established to exceed OSHA regulations applicable to the site.

Inhalation hazards are addressed adequately through Engineering Controls, as well as through Respiratory Protection and Personal Protective Equipment programs.

RECOMMENDATIONS FOR IMPROVEMENT

1. Evaluate the plant Industrial Hygiene Program to better align with current plant needs.

RESPONSIBILITY: Z. Vunderkind

PRIORITY: M

EDC: 12/31/2009

5.3 PRE-USE ANALYSIS

ACTUAL PERFORMANCE/CURRENT YEAR ACTIVITIES

Pre-use Analysis occurs through the following mechanisms:

- Management of Change (MOC) Procedure
- New Chemical Use Approval Procedure
- Loss Prevention reviews
- Capital Project Management (CPM) process

The plant's Management of Change program requires that any modifications to operating facilities undergo extensive safety review and approval, training, and documentation. A Request for Change Authorization (RCA) Form is used as a checklist to ensure that Management of Change procedures are followed, including Training, SOP revision, and Pre-Startup Safety Review, including verification that all safety concerns have been resolved prior to startup. Plantwide MOC awareness training is provided and audits are conducted periodically to verify compliance. This procedure was updated in 2008 and training will take place in 2009.

The New Chemical Use Approval Procedure covers the introduction of new plant chemicals or usages. It is referenced within the larger Management of Change procedure and checklist. This procedure helps ensure proper Hazard Assessment (including PPE) for new chemicals.

Process Hazards Analyses/safety reviews and Loss Prevention Control (LP) reviews are completed as regular steps in the Capital Project Management Process. These reviews involve a broad representation of plant personnel including operators and mechanics. Action items are documented and tracked to completion by the project engineer.

A review of Department RCA Logs indicates that approximately (40) modifications were initiated through MOC in 2008.

EFFECTIVENESS OF CURRENT PROGRAM

Management of Change has been effective in its purpose of identifying and safely implementing process changes.

The New Chemical Use Procedure has been effective when instituting use of major plant chemicals. Emphasis on use for minor chemical additions/changes is needed.

The Capital Project Management process has been very effective in managing safety and health-related aspects of engineering projects. Project reviews in their various forms are generally recognized as strong contributors to a project's safety and operability.

RECOMMENDATIONS FOR IMPROVEMENT

1. Provide MOC/New Chemical Use Procedure refresher training for supervisors and technical personnel.

ISSUED TO: J. King

PRIORITY: M

EDC: 12/31/2003

5.4 HAZARD ANALYSIS

ACTUAL PERFORMANCE/CURRENT YEAR ACTIVITIES

Job Hazard Analyses are addressed through the following programs:

- Job Safety Analysis (JSA)
- WE-R-SAFE. safety observation process
- Permitted procedures
- Contractor Safety Procedure/Permit

The plant JSA procedure specifies the purpose and process for performing Job Safety Analyses (JSA). In 2008, (49) archived JSAs were added to the on-line JSA database, which was created in 2006 to maintain JSAs for pre-job review and updated as needed. Three (3) new JSAs were also completed in 2008.

Permitted procedures include:

- Confined Space Entry—revised in 2007
- Hot Work—revised in 2007
- Contractor Safety—revised in 2006
- Impairment of Fire Protection Systems
- Impairment of Safety and Environmental Control Devices

Implementation of the behavior-based WE-R-SAFE. process was initiated in 2004 with development of an inventory of observation based on previous accidents. Observation data for 2008 compared to earlier years was as follows:

	2008	2007	2006
Observations	403	307	239
% Safe Observed	94	91	91
No. of Volunteer Observers	35	44	34

There was a decrease in the number of observers due to turnover of personnel at the site. As a result of WE-R-SAFE. observations, 46 action plans were developed with 37 completed. Our action plan for improving the process includes, but is not limited to, the following elements:

- Conduct observer coaching
- Observer Network Communications
- Conduct Observer training courses
- Implement the WE-R-SAFE. Corrective Action Plan process

In 2008, (1) Observer Training Course as well as plant-wide refresher training was conducted, and at year-end 100% of the plant workforce had been trained in the observation process. The WE-R-SAFE. process is believed to be a strong contributor to safety performance for both employees and contractors.

Job Hazard Analysis is a fundamental part of the Contractor Safety Procedure/Permit, and includes both Pre-Job Hazard Analysis and Construction Site Inspection checklists.

The process hazard analysis program provides for comprehensive review of all operating areas every (5) years. These reviews include key department supervisory, technical, maintenance, and operating personnel. Corrective Action Requests are tracked in accordance with the plant's SHCAR tracking process.

EFFECTIVENESS OF CURRENT PROGRAM

As a follow-up to two serious incidents in the past few years, the JSA program was revised to require JSAs for new and non-routine tasks, and to require trained observers to participate in JSAs. Implementation of these requirements took place 2008. The revision also included the WE-R-SAFE. process as the primary JSA mechanism.

Effectiveness of the WE-R-SAFE. observation process in identifying at-risk behaviors and addressing barriers to safe behavior continued to be very high in 2008, due largely to a 33% increase in the number of observations and improved corrective action follow-up.

Audits of permitted procedures in 2008 identified several areas for improvement which will be addressed as audit Corrective Action Requests. Overall, they are considered to be highly effective and are reviewed periodically.

RECOMMENDATIONS FOR IMPROVEMENT

1. Include audit of revised JSA procedure in 2009 S&H Audit Schedule.

ISSUED TO: J. King

PRIORITY: M

EDC: 12/31/2009

5.5 ROUTINE INSPECTIONS

ACTUAL PERFORMANCE/CURRENT YEAR ACTIVITIES

Routine inspection programs include the following:

- Safety & Housekeeping Inspections
- Various plant Mechanical Integrity Procedures

Monthly Plant-wide Safety and Housekeeping inspections of the entire site are coordinated by the PSC Vice-Chairperson and are conducted by teams which include ASC and PSC members and supervisors as well as other plant volunteers; Safety Department personnel participate in these inspections as well. The written inspection program includes a detailed safety inspection checklist for use as a guide. Findings are formally communicated in a written format to department supervision for resolution and response to the inspection team. Findings from previous inspections are checked during later inspections to confirm closure or work in progress, and repeat findings are noted as such on follow-up inspections. A training program on how to conduct inspections, including most common safety-related concerns, is provided to ASC and PSC members.

Formal department and plant inspection programs are in place for:

- Ladders, scaffolds, powered aerial platforms — revised in 2000 to require inspections for fixed ladders based on an incident
- Hoses
- Personal protective equipment
- Static grounding

- Vehicles
- Emergency safety showers and eyewashes
- Respiratory Protection Equipment

EFFECTIVENESS OF CURRENT PROGRAM

Effectiveness of the plant Safety and Housekeeping inspection program improved in 2008. A recommitment to completing inspections by increasing the emphasis on accountability for conducting and follow-up of safety and housekeeping inspections are primary contributors to this improvement. Continued improvement is needed in this area.

There are many formalized self-inspection programs addressed by S&H Procedures and effectiveness is generally good.

RECOMMENDATIONS FOR IMPROVEMENT

1. Improve accountability for completing monthly safety and housekeeping inspections and follow-up per established schedule.

ISSUED TO: I. M. Happy

PRIORITY: M

EDC: 12/31/09

5.6 EMPLOYEE HAZARD REPORTNG PROGRAM

ACTUAL PERFORMANCE/CURRENT YEAR ACTIVITIES

Employees may voice their concerns through many formal and informal channels, including direct notification of their supervisor. Employees recommend and lead JSA's, participate in incident investigations, participate in the safety observation process and department safety meetings, and offer input on safety reviews. Concerns can also be brought forth through the Plant Safety Committee. It is the intent that all issues be brought to closure and the employee originating the concern should be involved as much as possible. Members of the plant's Safety/IH department are available for consultation.

The formal program for addressing employee safety and health concerns is described in the Safety Concern Tracking and Follow-up Procedure.

Employee and contractor safety concerns (including anonymous concerns) are submitted, addressed, and tracked as Hazard Correction Requests (HCRs). An on-line system is available for submitting HCRs electronically, and anonymously if desired. The HCR database, which is maintained by the Safety and Health Manager, is incorporated into the plant-wide on-line SHCAR tracking system, under the HCR.

In addition, WE-R-SAFE has a corporate hotline for reporting of S&H concerns which do not appear to be addressed at the site level.

EFFECTIVENESS OF CURRENT PROGRAM

The HCR system has been very effective in capturing and tracking employee safety concerns, as indicated by the following summary:

	2008	2007	2006
HCRs submitted	7	20	22
HCRs completed	11	16	25

The drop in HCRs submitted in 2008 is considered to be caused largely to improved effectiveness of other measures, in particular the monthly plant visits instituted by the PSC, where concerns are actively captured.

RECOMMENDATIONS FOR IMPROVEMENT

None.

5.7 ACCIDENT/INCIDENT INVESTIGATIONS

ACTUAL PERFORMANCE/CURRENT YEAR ACTIVITIES

The plant's Incident Investigation Program covers investigation of all Unusual Incidents, including injuries, illnesses, and process-related incidents. All significant incidents, including Recordable injuries and illnesses, fires, and reportable spills and releases, are formally investigated by a team. Contractor injuries are investigated by the plant if plant personnel were involved, otherwise the contractor is requested to investigate and provide the plant with a report. Significant near-hits (i.e. near-misses) are formally investigated upon request. Informal investigations on first-aid injuries and near-hits are generally conducted by department personnel. In 2008, (10) formal incident investigations were conducted, (2) of which were of Near-Hits.

Each formal investigation is chaired by a member of the S&H staff, who is trained and experienced in incident investigation, and always includes the department First- or Second-Line Supervisor, other members of the department (including those involved in the incident), and usually a representative from another department. All incident investigations include at least one hourly or non-exempt employee. A WE-R-SAFE. Steering Team member is included when at-risk behavior is believed to have been a major contributing cause. Incident investigations are focused on identifying and addressing root cause(s), not on assessing blame. Incident-related disciplinary issues are handled outside of the incident investigation process.

Each investigation report includes a chronologic description of the incident, primary and contributing causes, and corrective actions to prevent recurrence. A primary cause classification is assigned, such as mechanical failure, procedural inadequacy, human error, at-risk behavior, or unsafe condition. A behavioral analysis is also conducted where appropriate. S&H Corrective Action Requests are issued to address significant findings. The owning department is responsible for follow-up on informal investigations.

After review and approval by the ASC, incident reports are distributed plant-wide and are reviewed in department safety meetings.

Incident follow-up is tracked in accordance with the S&H Corrective Action Tracking. The SHCAR database is used to track S&H Corrective Action Requests from formal incident investigations. In 2008, (17) incident investigation SHCARs were initiated and (42) completed.

A process for capturing and communicating near-hit incidents, including anonymous on-line reporting, was implemented in 1999, and as a result the number of near-hits reported is as follows:

	2008	2007	2006
Near Hits Reported	18	33	18

In 2002, emphasis was placed on follow-up from near-hit reports; as a result 90% of corrective actions from near-hit reports were completed.

EFFECTIVENESS OF CURRENT PROGRAM

The current incident investigation program is considered to be very effective. Incidents are effectively investigated by knowledgeable, experienced personnel. Findings are tracked to completion via the on-line SHCAR Tracking Database. Emphasis on improving reporting and follow-up of near-hit reports is on-going.

RECOMMENDATIONS FOR IMPROVEMENT

None.

5.8 TREND ANALYSIS

ACTUAL PERFORMANCE/CURRENT YEAR ACTIVITIES

Trend analysis is performed routinely as part of the WE-R-SAFE. behavioral safety observation process. Data from observations on at-risk vs. safe behaviors are entered into a computer program which tabulates percent safe and at-risk for each critical behavior listed. This data is used to identify focus areas for improvement, primarily via action plans initiated by the WE-R-SAFE. Steering Team.

For 2008, overall percent safe behaviors observed improved as follows:

	2008	2007	2006
%Safe	94	91	91

Top (5) at-risk (lowest %safe) behaviors observed were as follows:

	%Safe		
Category	2008	2007	2006
1. Fall protection	65	60	41*
2. Lifting/bending/twisting	78	74	77
3. Loading Racks	79	80	75
4. Awkward/Cramped	82	83	89
5. Ascending/Descending	85	87	93

All of the behaviors listed above will be reviewed by the appropriate person(s) to determine how to increase the safe behaviors associated with these items as they apply to each area of the plant. This will be done through our WE-R-SAFE Corrective Action Process.

*As per the chart above, Fall Protection % safe has increased not only because modifications and repairs were performed on some loading racks, but there is an increased understanding and recognition of risks associated with the use of fall protection overall. This will remain a focus area considering the severity of injury related to falling potentials.

Another focus area will be Ascending/Descending as it is noted that this behavior is increasingly becoming more At-risk.

As two of the three Recordable incidents in 2008 involved walking-working surfaces, a plant-wide inspection was initiated to identify and correct any other problem areas. Because the number of incidents is typically low, trend analysis on upstream behavioral data is much preferred over downstream incident data. Injury reports are reviewed annually by the WE-R-SAFE. Steering Team to identify behavioral trends.

EFFECTIVENESS OF CURRENT PROGRAM

The WE-R-SAFE. process provides an on-going, proactive, analytical approach to trend analysis of upstream at-risk behaviors, which can then be addressed via Action Plans. This is believed to be the best way to identify potential focus areas for injury prevention efforts.

Recognition and reporting of occupational illnesses has improved over the past few years. Continued emphasis will provide focus areas for illness-prevention programs (primarily ergonomics).

RECOMMENDATIONS FOR IMPROVEMENT

None.

6.0 HAZARD PREVENTION AND CONTROL

6.1 CERTIFIED PROFESSIONAL RESOURCES

ACTUAL PERFORMANCE/CURRENT YEAR ACTIVITIES

The plant Safety and Health Manager is a Certified Safety Professioal with over 20 years experience in manufacturing and engineering, including over 10 years in safety. The plant Industrial Hygienist is a Certified Industrial Hygienist (CIH). The staff Safety/IH Engineer is an Associate Safety Professional (ASP) with a degree in Environmental Health. In addition, CIHs and CSPs are available from consultants for audits and consultation on an as-needed basis.

EFFECTIVENESS OF CURRENT PROGRAM

The amount of expertise available within the plant, through corporate support groups and outside consultants continues to be adequate for the facility.

RECOMMENDATIONS FOR IMPROVEMENT

None.

6.2 HAZARD PREVENTION AND CONTROL

ACTUAL PERFORMANCE/CURRENT YEAR ACTIVITIES

Recognized safety and health hazards are eliminated or controlled in a variety of ways. The primary mechanisms for accomplishing this goal at the WE-R-SAFE are:

- Engineering Controls
- Adminstrative Controls
- Safety and Health Rules
- Personal Protective Equipment
- Hazard Control Programs

A. Engineering Controls

Local exhaust systems designed for personnel exposure reduction are provided in the following areas:

- Welding
- Product Packaging
- Shop machine tools
- Shop benchtop ventilation (pump repair area)

Local exhaust ventilation systems are checked on an established frequency (at least annually) to ensure that performance is satisfactory. A ventilation system testing procedure is in place to document this program.

In 2008, there were several upgrades to the Waste Treatment Plant control room to reduce nuisance vapors and odors from the process control area, including separating the lab area and providing local exhaust for a sample oven. There was also a new hydrochloric acid loading rack installed that eliminated the need for a personal fall arrest system due to the presence of primary fall protection.

B. Administrative Controls

Administrative controls employed primarily include safe work permits and training. Employee job rotation is not used as an administrative control. Permits and Training are discussed in separate sections.

C. Safety and Health Rules

Most of the plant's Safety and Health Rules are contained in the written safety and health program which consists of four manuals, which together include over 50 S&H procedures, and the Guidelines for Contractors handbooks. Department-specific safety and health rules for operating units are included in their Standard Operating Procedures.

Basic safety and security rules are included in the new employee safety orientation and training, as well as orientations for contractors and visitors/drivers/service personnel.

D. Personal Protective Equipment

Personal protective equipment programs are covered primarily by the following:

- Respiratory Protection Program

- Personal Protective Equipment
- Eye and Face Protection

Other S&H procedures which specify PPE include:

- Hearing Conservation
- Fall Protection
- Breaking into Process Equipment
- Emergency Response Plan
- Contractor Work Permit

The Respiratory Protection program includes all required elements, including selection, training, fit-testing, medical surveillance, and program evaluation. An internal audit of the program in 2008 revealed a high level of compliance and only minor deficiencies.

The Personal Protective Equipment procedure includes all necessary elements, including hazard assessment, selection, storage, training on use and limitations, cleaning/decontamination, inspection of PPE, and approval prior to purchasing equipment.

An on-line computer-based PPE Reference Guide is available for all employees, which lists for each major material the following:

- Physical description
- Physical Properties

- Exposure Limits/Guidelines
- Health effects, acute and chronic
- PPE Requirements specific to tasks and by generic category

Eye and Face Protection provides information as to what safety glasses and goggles are acceptable in the plant.

All costs for uniforms, safety glasses and other PPE are paid for by WE-R-SAFE.

E. Hazard Control Programs

Hazard Control Programs are in place which include:

- Preventive Maintenance
- Hazard Correction Tracking
- Occupational Healthcare Program
- Disciplinary Program
- Emergency Response Plan

See appropriate sections for a review of these elements.

EFFECTIVENESS OF CURRENT PROGRAM

A. Engineering Controls

Overall effectiveness of engineering controls are generally considered to be adequate. Continued work on reducing fugitive emissions sources is expected to further reduce chemical exposures in the plant.

B. Administrative Controls

Overall effectiveness of administrative controls is considered to be adequate at this time.

C. Safety and Health Rules

Overall effectiveness of Safety and Health Rules is believed to be adequate at this time.

D. PPE

The overall effectiveness of the Personal Protective Equipment procedures appears to be much improved. There has been a strong emphasis on PPE use, which will continue to be evaluated as part of the WE-R-SAFE. process. PPE, including respiratory protection, requires frequent monitoring to reinforce the need for use.

The new on-line PPE database is an excellent reference tool for chemical protective measures; in 2008 plant-wide training was conducted on PPE including how to use the on-line system.

E. Hazard Control Programs

Overall effectiveness of Hazard Control Programs is believed to be adequate at this time.

RECOMMENDATIONS FOR IMPROVEMENT

None.

6.3 PROCESS SAFETY MANAGEMENT

We-R-Safe does not use or store any hazardous chemicals in quantities above the Threshold Quantity that would place it under the OSHA PSM standard (29CFR1910.119)

6.4 PREVENTIVE MAINTENANCE

ACTUAL PERFORMANCE/CURRENT YEAR ACTIVITIES

The preventive maintenance program is covered by a number of Mechanical Integrity Procedures. These include:

- S&H Equipment
- Tank and Vessel Inspection
- Fire Protection Equipment Inspection and Maintenance
- Testing of Safety/Environmental Control Devices
- Static Grounding of Equipment and Buildings
- Maintenance of Safety Relief Devices

The plant's Preventive Maintenance (PM) program is managed using the SAP Maintenance Module. Work orders for tasks such as inspection, testing, calibration, lubrication and replacement are automatically generated at specified frequencies for a significant amount of the plant's operating equipment. Inspection and testing frequencies are determined by applicable engineering guidelines or by past equipment history. Separate inspection programs for tanks, vessels, and piping are maintained by the plant Reliability Engineer who is ASME-certified.

Reliability-based maintenance processes, including Asset Effectiveness Management (AEM) and failure analysis, and reliability-centered maintenance are used to improve mechanical and safety reliability.

The preventive maintenance program is audited annually as a part of the plant's Risk Management Program self-audit in accordance with State PSM.

EFFECTIVENESS OF CURRENT PROGRAM

During 2008, The plant upgraded to a new version of the SAP Maintenance Module, which it uses to manage the PM program. It was recently discovered that not all of the information in the old system was properly transferred to the new system. A comprehensive review of all safety-critical equipment should be completed to ensure that they are all appropriately included in the system. Upon completion of this review, PM program management responsibilities should be reaffirmed/reassigned to ensure continued effectiveness.

It should be noted that much of the plant's improvement in reducing environmental releases over the years can be attributed to fewer mechanical failures of equipment.

RECOMMENDATIONS FOR IMPROVEMENT

1. Conduct comprehensive review of safety-critical equipment to ensure that all equipment is appropriately included in the SAP PM program.

ISSUED TO: C. DeWinner

PRIORITY: H

EDC: 12/31/09

2. Reaffirm/reassign as needed responsibilities for management of the plant Preventive Maintenance Program.

ISSUED TO: C. DeWinner

PRIORITY: H

EDC: 12/31/09

6.5 HAZARD CORRECTION TRACKING

ACTUAL PERFORMANCE/CURRENT YEAR ACTIVITIES

The primary program for tracking safety and health concerns/findings is the plant's S&H Corrective Action Request (SHCAR) database, described in the S&H Corrective Action Tracking and Follow-up. This system tracks Corrective Actions from the following sources:

- Incident Investigations
- Internal Audits
- Employee Safety Concerns
- Safety and Health Management Program Self-Evaluations

This system is tied to the plant e-mail system and automatically generates SHCAR look ahead reminders and overdue notices, as well as monthly summary of Key Performance Indicators (#Open, #Completed YTD, #Generated YTD, #Past Due, and %Past Due) which is monitored by the ASC.

Employee and contractor safety concerns (including anonymous concerns) are submitted, addressed, and tracked in the SHCAR database as Hazard Correction Requests (HCRs) in accordance with the Safety Concern Tracking and Follow-up Procedure.

Action items which arise from near-hits are tracked in the Near-Hit Database which is periodically updated by the WE-R-SAFE. coordinator. All actionable items from behavior based safety observations are also tracked by the WE-R-SAFE coordinator.

Findings from monthly Safety and Housekeeping Inspections are tracked by the monthly inspection teams to verify completion. The Plant Safety Committee also tracks correction of safety concerns through monthly meeting minutes.

EFFECTIVENESS OF CURRENT PROGRAM

The current processes for hazard correction tracking are deemed adequate. In 2008, a record (173) SHCARs were completed. This system is considered to be a strong contributor to organizing safety improvement efforts in the plant.

RECOMMENDATIONS FOR IMPROVEMENT

None.

6.6 OCCUPATIONAL HEALTHCARE PROGRAM

ACTUAL PERFORMANCE/CURRENT YEAR ACTIVITIES

For routine medical services, the WE-R-SAFE retains the services of a Certified Occupational Healthcare Physician. That doctor provides essentially all occupational medical services required under normal conditions including: pre-placement and periodic physical examinations, OSHA required medical evaluations and medical care beyond first aid (Medical transport to a local hospital would be utilized for more serious injury cases). Supporting documents for Respiratory Protection, HAZWOPER, Hearing Conservation and Bloodborne Pathogens are provided to the doctor by the plant Industrial Hygienist.

The Industrial Hygiene Status Report and Assessment Strategy is periodically reviewed with the doctors assistance and a copy provided. This report summarizes employee exposure monitoring and assessments conducted for the major materials used and produced at the facility. The doctor has also toured the WE-R-SAFE to become familiar with the facility, materials used and produced and the job activities of varying job classes.

The plant maintains a first aid room with supplies for treatment of minor injuries. Several employees are designated first aid providers and are required to maintain basic certifications in first aid, CPR, and Bloodborne Pathogens certifications. Around the clock coverage is provided for First Aid/CPR. The plant also has an Automatic External Defibrillator (AED) for use by plant first aid personnel. A new trainer must be identified for First Aid/CPR as the plant's certified trainer is no longer available.

EFFECTIVENESS OF CURRENT PROGRAM

The current Occupational Healthcare Program has been in place for several years and is deemed to be effective for meeting plant needs. The first aid refresher training schedule, including AED use, should be reviewed and modified to ensure adequacy.

RECOMMENDATIONS FOR IMPROVEMENT

1. Identify new First Aid/CPR trainer and review/modify the first aid training schedule, including AED use, to ensure adequacy.

ISSUED TO: J. King

PRIORITY: H

EDC: 12/31/09

6.7 DISCIPLINARY PROGRAM

ACTUAL PERFORMANCE/CURRENT YEAR ACTIVITIES

While supervision is primarily responsible for enforcement and reinforcement of safety and health requirements, all employees are expected to do their part to help ensure that safety and health requirements are met by both employees and contractors.

Job safety performance is considered part of each person's total job performance for both salaried and hourly personnel, and poor performance is subject to disciplinary action up to and including termination.

In 2008, the plant's Policies and Procedures Manual, which includes a section on Discipline, was updated and communicated to all employees. It calls for progressive discipline for work performance problems, including safety-related.

In 2008, WE-R-SAFE developed and issued a list of Cardinal Safety Rules, which are considered to be grounds for termination if violated.

EFFECTIVENESS OF CURRENT PROGRAM

The current disciplinary system that is in place is considered to be adequate, however there has seldom been a need to use it for safety-related performance problems.

RECOMMENDATIONS FOR IMPROVEMENT

None.

6.8 EMERGENCY PROCEDURES

ACTUAL PERFORMANCE/CURRENT YEAR ACTIVITIES

The plant has a comprehensive Emergency Response Plan which establishes responsibilities, procedures, and equipment for potential emergency situations, as well as training, drill and audit requirements. This plan was completely reviewed, updated and reissued during 2008.

Two plant emergency drills were conducted in 2008, including an annual plant evacuation drill and a fire/rescue drill involving outside fire and EMS companies.

All drills and response to actual incidents are critiqued and summarized in drill and incident reports, with items for improvement tracked in accordance with the S&H Corrective Action Request process.

During 2008, around the clock guard coverage was instituted due to security concerns, which also improved off-shift emergency response capability.

EFFECTIVENESS OF CURRENT PROGRAM

The plant's Emergency Response Plan addresses all appropriate issues and is believed to be adequate.

RECOMMENDATIONS FOR IMPROVEMENT

None

7.0 SAFETY AND HEALTH TRAINING

7.1 SUPERVISOR SAFETY AND HEALTH TRAINING

ACTUAL PERFORMANCE/CURRENT YEAR ACTIVITIES

Supervisors participate in the safety and health training program according to their responsibilities. Supervisors are required to review Computer-Based Training modules and attend plant-wide training sessions that are pertinent to their job. S&H-related training and communication also takes place at monthly supervisor meetings conducted by the Plant Manager.

Training for new supervisors is primarily on-the-job and includes a review of safety and health requirements and responsibilities that affect their area. A Personal Safety Checklist for Supervisory Personnel is used for this purpose.

EFFECTIVENESS OF CURRENT PROGRAM

Supervisors have the training and experience that they need to effectively carry out their safety and health-related responsibilities.

RECOMMENDATIONS FOR IMPROVEMENT

None.

7.2 EMPLOYEE SAFETY AND HEALTH TRAINING

ACTUAL PERFORMANCE/CURRENT YEAR ACTIVITIES

The plant training effort is coordinated by the Plant Training Team which includes the Plant and Department Training Coordinators as well as S&H staff. Initial/Refresher S&H Training Requirements Matrices are used to develop annual training plans and new employee safety training schedules. Training is tracked using a proprietary software program.

The following plant-wide S&H refresher training was provided in 2008:

- Hazard Communication-General Awareness (CBT)
- Hearing Conservation (CBT)
- Respiratory Protection (CBT)
- Emergency Response-Awareness Level (CBT)
- Emergency Response-Operations Level
- DOT General Awareness (CBT)
- Portable Fire Extinguishers (CBT)
- Fire Protection Systems/Impairments (CBT)
- Lockout/Tagout (CBT)
- PPE Awareness (CBT)
- PPE Operations
- Job Safety Analysis
- WE-R-SAFE. Overview

Practically all initial and refresher training includes written verification of understanding, with 80% established as a passing score. Certifications for Operator and Emergency Response training are provided.

Department safety meetings are also used for department-specific training topics.

EFFECTIVENESS OF CURRENT PROGRAM

The overall level of effectiveness of plant training efforts is very high, in particular systems to plan, track, deliver, conduct and verify learning. Completion of mandated S&H training in 2001 was 100%. Use of Authorware CBT for routine Departmental refresher training is planned to enhance efficiency of this training.

RECOMMENDATIONS FOR IMPROVEMENT

None.

7.3 EMERGENCY RESPONSE TRAINING

ACTUAL PERFORMANCE/CURRENT YEAR ACTIVITIES

Emergency response training conducted in 2008 included:

- Confined Space Rescue training was provided in accordance with the OSHA standard.

- Semi-annual Incident Command refresher training for Chief Coordinators which include discussion of table-top scenarios. Only one session was conducted in 2008.
- Operations-level emergency response refresher training for all operations personnel, which includes discussing table-top scenarios.
- Awareness-level emergency response refresher training for all plant personnel via computer-based training.
- Training to maintain first aid/CPR certifications for laboratory personnel.

Department-specific emergency response training in most areas.

EFFECTIVENESS OF CURRENT PROGRAM

Current levels of emergency response training are adequate to meet existing needs.

RECOMMENDATIONS FOR IMPROVEMENT

None

7.4 PPE TRAINING

ACTUAL PERFORMANCE/CURRENT YEAR ACTIVITIES

The plant's PPE training program includes initial and periodic refresher training in several areas:

- Awareness Level (all employees-annual)— basic head, eye, and foot protection

- Operations Level (operations/maintenance employees-annual)—chemical protective clothing, gloves)
- Respiratory Protection (operations/maintenance/emergency response personnel-annual)
- Hearing Protection (based on potential for exposure-annual)
- Fall Protection (operations/maintenance employees)
- Departmental Operator Training programs

PPE requirements for contractors and visitors are covered as part of their respective safety orientation programs.

EFFECTIVENESS OF CURRENT PROGRAM

Current PPE training is considered to be adequate. WE-R-SAFE. observations indicate an improvement in %Safe with regard to PPE improved from 91% in 2007 to 94% in 2008. Emphasis on reducing at-risk behavior in this area is on-going through the WE-R-SAFE. observation process and supervisor reinforcement.

RECOMMENDATIONS FOR IMPROVEMENT

None.

7.5 MANAGER SAFETY AND HEALTH TRAINING

ACTUAL PERFORMANCE/CURRENT YEAR ACTIVITIES

Managers participate in the same safety and health training programs as all employees, according to their responsibilities. Managers are required to review Computer-Based Training modules and attend initial safety orientation and plant-wide training sessions that are pertinent to their job. S&H-related training and communication also takes place at monthly supervisor meetings as well as at ASC meetings.

EFFECTIVENESS OF CURRENT PROGRAM

The existing training program for managers is deemed effective.

RECOMMENDATIONS FOR IMPROVEMENT

None.

8.0 SIGNIFICANT CHANGES IN SAFETY & HEALTH MANAGEMENT SYSTEM

The following significant changes were made to the safety and health management system in 2008:

- Elimination of OLDCOMPANY Contractor/Guest S&H Guidelines— these were largely replaced by a new plant Contractor Approval and Selection Procedure and updated Contractor Safety Procedure which provide a contractor safety management process more appropriate to the site.
- The plant converted from OLDCOMPANY's SAP system to WE-R-SAFE's in February, 2008, which impacted the Preventive Maintenance program

9.0 SUCCESS STORIES

- For the second year in a row, the site received a Safety Award from the County Office of Emergency Management for our continuing good performance and commitment to safety.

- Attachment 6

Sample Job Hazard Analysis

Job Hazard Analysis

Title of Task: Staging material for processing		
Department: Manufacturing		
Author of JHA: Tom Jones	Date of orig.: 3/10/05	Date of rev.: 10/24/08
Reviewed: Norman Lynch	Approved: Brian Smith	
PPE Required: Kevlar ® Gloves, hard hat, safety shoes		
Steps of Task	Hazards of Step	Preventive Measures
Lift a 50lb box from a pallet on the floor	Back strain from lifting a heavy load	Place pallets on self-leveling platforms to maintain a proper height of the material
Place the box on the work table to the side	Strain from twisting under load	Relocate the worktable to the front to eliminate any twisting
Open the box with a box cutter	Cuts from the exposed blade of the box cutter	Use self retracting box cutters and Kevlar® gloves to prevent cuts
Remove the material and place it on the table	Arm strain from lifting the material out of the box	Use a tilting table or a vacuum lift to eliminate the stretching and reaching motion

Attachment 7

Work-Related Incident Report Form

We-R-Safe

Unsafe Condition/Near Miss/Incident/

Safety Suggestion Report

CHECK ONE	UNSAFE CONDITION	NEAR MISS	SAFETY SUGGESTION
LACK OF LOCKOUT/TAG OUT	WORKING UNSAFE W/EQUIPMENT	UNSAFE AREA	DEFECTIVE EQUIPMENT
RUNNING	PROCEDURES NOT FOLLOWED	LACK OF GUARD	LACK OF LIGHTING
NOT USING PPE	SAFETY RULES NOT FOLLOWED	ELECTRICAL HAZARD	SAFETY DEVICES INOPERATIVE
NOT USING FALL PROTECTION	LACK OF KNOWLEDGE	CHEMICAL HAZARD	POOR HOUSEKEEPING
GUARD WAS BYPASSED/RE MOVED	____________	FIRE HAZARD	____________

Date ________________ Time ________________

Department________________________

Location____

Name (optional)____

Explain Condition____

Severity Rating

Note—This form is not to be used For 'Catastrophic' Severity—if this severe of a condition is found contact your supervisor immediately

1- Critical **(may cause severe injury, illness, lost workdays, or major property damage)**

2- Marginal **(may cause minor injury, illness or minor property damage)**

3- Negligible **(probably would not affect safety/health, but is a safety violation)**

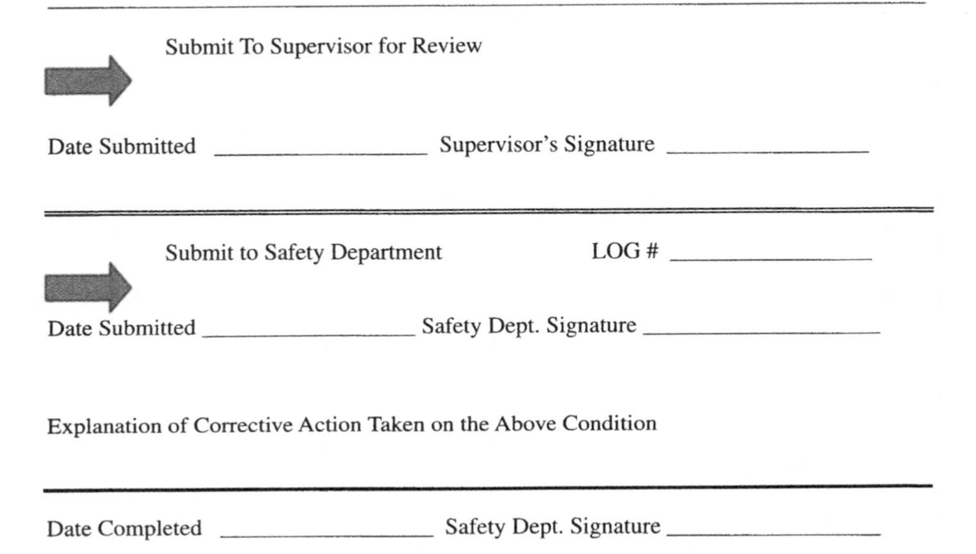

Submit To Supervisor for Review

Date Submitted ________________ Supervisor's Signature ________________

Submit to Safety Department LOG # ________________

Date Submitted ________________ Safety Dept. Signature ________________

Explanation of Corrective Action Taken on the Above Condition

Date Completed ________________ Safety Dept. Signature ________________

4

MANAGEMENT LEADERSHIP AND EMPLOYEE INVOLVEMENT

INTRODUCTION

The first, and arguably the most important of the four VPP cornerstones, is management leadership and employee involvement. A safety and health management system, regardless of how comprehensive and complete, must have the full commitment and support of management. Managers must provide the leadership to guide the organization to success. Employees must fully embrace and actively participate in all aspects of the safety and health process in order for it to be successful and ultimately achieve VPP recognition.

Traditionally, safety and health was a program that was implemented over a period of time with the hopes it would achieve a specific goal, such as reducing injuries, achieving regulatory compliance, or lowering workers' compensation costs. Traditional safety programs were management driven from the top down, and employees were expected to blindly and obediently follow along or else. The "Safety God" made the rules and handed them down to the workers for implementation without any input as to their practicality or applicability. Management communicated throughout the organization that safety and health was now a priority, and heaven help those who did not follow the rules and achieve the goals. Safety and health was managed by fear, not by inclusiveness. Some may believe that this approach was, and still is, effective. However, employees must believe and willingly participate in the safety and health process in order for it to succeed. A safety and health system is much more effective if employees participate because they want to rather than they have to out of fear.

Preparing for OSHA's Voluntary Protection Programs. By Brian T. Bennett and Norman R. Deitch
Copyright © 2010 John Wiley & Sons, Inc.

As draconian as this process was, there may have been some minimal, incremental improvement in safety and health performance over time, but was safety excellence really achieved? Were the short-term improvements achieved real lasting, systemic improvements or just luck? Did we achieve everything that we really could have had things been a little bit different? The answer, not surprisingly, is probably not.

The fact is the more each and every employee is involved in the safety and health process, and the greater their involvement, the more likely they will feel somewhat responsible for ensuring the success of the overall safety and health process.

Employees are much more likely to develop a sense of ownership for the safety and health process and support its requirements without reservation if they not only implement it but also help to develop and maintain the various programs. Employees at VPP participants have actually expressed their involvement in the safety and health process as a sense of ownership. Management should provide the necessary leadership to encourage employees from all levels and all departments of the organization to become involved in all aspects of the safety and health process.

Safety and health is not just for hourly employees or first-line supervisors; it is for everyone, from top to bottom. Well-written safety and health systems have failed because management did not adequately demonstrate to all employees that they were serious about safety and health, and that they were committed to make sure the safety and health process was fully implemented. There are a number of ways management can demonstrate their commitment to the safety and health process:

- Management should routinely inspect the work site, both in a formal and informal manner. During the VPP onsite evaluation interviews, employees should be able to report that they frequently observe management in the workplace.
- Safety and health rules apply to all employees. Does management wear all required personal protective equipment when in the workplace or are the rules just for the workers?
- Does management participate in the various safety and health project teams, committees, work groups, and the like as active members? Do they regularly attend the meetings and contribute to the success of the team? Do they accept action items and complete them in a timely manner?
- Does management attend safety and health training sessions as required? Do they show up late or leave early? Do they participate?
- Does management present safety and health training sessions to employees? Are they on time? Are they adequately prepared and familiar with their topic of instruction?
- Does management informally talk to employees about safety and health, to find out if there are any issues or concerns and to ensure there are no problems that may need their attention? Or, does management just talk to employees when there are problems? Does management ever publically recognize good performance?

MANAGEMENT LEADERSHIP

Management has the responsibility to set the tone, provide leadership, and demonstrate their commitment in support of the safety and health process. If management does not make safety and health a core value within the organization, VPP recognition is unlikely to be achieved regardless of the resources committed. Management must get out to the workplace and "walk the talk," and provide highly visible leadership for all employees to see.

Management must make sufficient resources available to ensure success, track the progress of the development of the safety and health process, and hold everyone accountable for their performance, ensuring that everyone has met all of their responsibilities in regards to safety and health. There is an old management adage: What gets measured gets done. If management does not put the appropriate systems in place to ensure everyone is doing their part and goals and objectives are being developed and met in a timely manner, employees will interpret that as management does not think safety and health is really important. If employee perceptions ever get to this point, the safety and health process is doomed to failure regardless of how well-intentioned management may be. It is therefore incumbent upon management to set and enforce the proper tone in regards to safety and health.

Management must demonstrate their leadership by being visibly active in all aspects of the safety and health process. Employees look to management to determine the culture of the organization. Employee perceptions of management's commitment will set the tone of the organization. If management cannot take the time to participate and support the safety and health system, employees will again perceive a lack of importance and will behave in much the same manner.

SAFETY AND HEALTH POLICY

The first place to start developing the management leadership cornerstone is to develop a safety and health policy. A concise, well-written safety and health policy is a very visible and oftentimes an employee's (and an outsider's) first exposure to an organization's safety and health culture. Quite often, an effective safety and health policy can be written in just a few sentences. The safety and health policy will set the tone for all subsequent safety and health activities.

It is not sufficient for management to just write and post a safety and health policy to result in a successful process. Employees are used to a host of programs and projects being implemented to enhance the business. Management must prove to all employees that safety and health is a core value, that they believe in the system, and they are committed to its success. Quite simply, management must walk the talk. Management must demonstrate to everyone that they support the safety and health process by being actively and visibly involved in all aspects of the process.

For large organizations, there should be an overarching corporate safety and health policy that sets the general policy and philosophy of the organization in terms of safety and health. The policy should be signed by the organization's chief executive officer as a sign of support. There should also be a written site-specific safety and health policy. Ideally, the safety and health policy should be jointly signed and issued by the senior management executive as well as the union president (for represented sites) or the hourly safety and health committee chair or representative at nonrepresented sites. By having both management and hourly representatives sign the policy, all employees are aware of the leadership and commitment to the safety and health process of each of the organization's internal stakeholders.

The safety and health policy is best developed by a multidiscipline committee of both management and hourly employees rather than an individual. Sufficient time must be allocated to ensure each of the elements deemed important by the organization are captured in the safety and health policy. The local safety and health policy should not only set the framework of the safety and health system, but should make note of the organizational and the signators' personal importance of safety and health. Key elements to include in the safety and health policy include:

- Statement of the importance of safety and health to the organization and the local site leadership
- Statement of safety and health responsibilities
- Statement of how employees will be held accountable

The safety and health policy should be reviewed periodically to ensure it reflects the current culture and issues at the site. If the responsibilities of the signatories change, the policy should be reviewed and modified as appropriate. Exhibit 4.1 is an example of a comprehensive site-specific safety and health policy.

Once the safety and health policy has been developed and signed, it must be communicated to all employees. This is important as all employees must understand the importance of safety and health to the organization. The safety and health policy must be communicated to all new employees as part of their new employee orientation. It is also a good idea to review the safety and health policy with all employees on an annual basis at a safety meeting or toolbox talk. The safety and health policy should be posted on bulletin boards so that it is highly visible for all employees to see. Some organizations have provided their employees with laminated wallet cards or hard-hat stickers of the policy so it is constantly in front of the employees. Another method for keeping the safety and health policy prominent is to integrate it on cards that employees must keep visible at all times.

Although there is no requirement for employees to memorize the safety and health policy, for sites undergoing a VPP on site evaluation, the expectation is that the employees will have a general understanding of what the safety and health policy says.

SAFETY AND HEALTH GOALS AND OBJECTIVES

Once the safety and health policy has been developed, safety and health goals and objectives must be developed to support it. As with the policy, safety and health goals and objectives should be developed by a multidisciplinary team involving both management and hourly employees. Goals and objectives should be developed for the overall site as well as for each individual department. Some organizations develop goals and objectives to achieve long-term issues, such as a 3- or 5-year plan. Individual department goals are usually short-term projects that support the long-term organizational goals. It is common and acceptable to find the annual safety and health goals incorporated in quality plans, operating plans, or business plans.

As with any goals, the following should be considered:

- *Simple* The VPP does not require complex goals and objectives. In fact, the simpler the better so long as achieving the goal will enhance employee safety and health.
- *Measureable* Progress toward achieving the goal as well as being able to measure the final outcome.
- *Achievable* Goals should be formatted such that they can be accomplished by the appropriate group (e.g., employee team, departmental team, site-wide committee, etc.)
- *Realistic* The goal should be realistically achievable in the allotted time.
- *Timely* The goal should address a current issue or problem.

The expectation for VPP sites is that there will be meaningful goals and objectives established annually to support and enhance the safety and health system. The safety and health goals and objectives should be in a narrative format. Numeric goals and objectives, such as zero injuries or accidents, with no supporting goals, are not considered adequate. A more appropriate goal would be to reduce injuries to maintenance mechanics by providing hand tool safety training and implementing a daily hand tool inspection program. Exhibit 4.2 is an example of long-term organizational goals and supporting short-term departmental goals.

Safety and health goals and objectives should be evaluated periodically to ensure progress is being made in completing them. The summary of safety goals and objectives progress can be included in various minutes of committees or work teams.

SAFETY AND HEALTH ORGANIZATION

The safety and health function must be integrated into the site's overall management structure. There is no requirement for a VPP work site to have a full-time safety and health professional on staff. In fact, many smaller VPP work sites may assign the safety and health responsibilities to employees as a collateral duty, or may spread out the safety and health responsibilities among several employees.

It is recommended that the person assigned safety and health responsibilities has those job requirements incorporated into his or her job description. The safety and health person should also receive some basic safety and health training in order to do his or her job properly. Individuals assigned safety and health responsibilities should be clearly identified on the site's organization chart. Lastly, the person assigned primary responsibility for safety and health should report to someone close to the top of the site's organization chart, preferably the site's chief executive. Having the person responsible for safety and health report to the chief executive ensures that safety and health is highly visible to upper management, and a person with a high level of authority within the organization can help resolve any issues or problems in a timely manner.

Your VPP application must include a site organization chart that illustrates the relationship of the safety and health function to your management organization. If there are several employees responsible for the site's safety and health functions, it is helpful to attach an organization chart of the safety and health functions. The organization may consist of safety professionals, industrial hygienists, technicians, ergonomists, physicians, nurses, and others.

One of the basic concepts of a successful safety and health process is that safety and health is a line responsibility. However, there are responsibilities for staff employees as well.

Staff members generally manages and oversees the safety and health process, lending their support when necessary. Typical safety and health responsibilities for staff employees generally include:

- Involvement in the development and approval of safety and health policies and procedures
- Involvement in the development and approval of safety and health goals and objectives
- Involvement in the development and approval of the safety and health training program
- Ensuring adequate resources, including funds, personnel, and equipment, for safety and health are available
- Implementation of the safety and health discipline and accountability system

Line supervisors are usually more directly involved with the actual implementation and enforcement of the safety and health policies and procedures as they are on the "front lines" along with the employees. Typically, this includes:

- Conducting employee safety and health training
- Inspecting the work site for unsafe acts and conditions
- Facilitating safety and health committees, work teams, meetings, and the like
- Participating in the development of job safety analysis and other hazard assessments
- Holding employees accountable for their safety and health performance

Whatever the safety and health responsibilities may be, they should be documented in the employees written job descriptions. The expectation at VPP sites is that all employees will have at least one safety and health responsibility included in their job descriptions regardless of their position. One representative job description, which illustrates the inclusion of safety and health responsibilities, should be attached to your VPP application.

SAFETY AND HEALTH PERFORMANCE APPRAISALS

The next step after safety and health responsibilities have been assigned is to ensure that they are accomplished. A formal accountability system must be established to review the performance of all managers and supervisors on at least an annual basis. The expectation for VPP work sites is that there will be a written, detailed performance appraisal form to evaluate an employee's safety and health responsibilities. This appraisal form should be detailed and address each of the employee's specific safety and health responsibilities. It should be completed at least annually, and the results should be considered when determining pay increases and promotional opportunities. A blank safety and health appraisal form must be attached to your VPP application. Exhibit 4.3 is an example of an employee safety and health performance appraisal form.

There is no VPP requirement for written safety and health performance appraisals for hourly employees. The reason is that typically there is not a similar formalized safety and health performance evaluation system for hourly employees. In fact, represented employees are usually not formally evaluated at all. However, there are informal appraisal systems that may be used for hourly employees that do not impact their wages, promotional opportunities, or result in other punitive measures. For nonrepresented employees, it is common to find written safety and health performance appraisals. Certainly, a verbal review between hourly employees and their supervisor on an annual basis to discuss safety and health performance and issues is a good practice.

SAFETY AND HEALTH RESOURCES

Management is responsible for ensuring that adequate resources are available for the development, implementation, and ongoing maintenance of an effective safety and health process. Safety and health resources include:

- Personnel
  - Management must ensure that a sufficient number of appropriately trained personnel are available to support the safety and health process commensurate with the risk at the work site. Among the type of personnel that may be required, based on the specific hazards and complexity of the workplace, include: safety and health professionals, industrial hygienists, health physicists, ergonomists, emergency responders, physicians, and nurses. As mentioned previously, there is no VPP requirement that any of the safety and health staff be full-time employees at a work site or even employees of the organization. The use of corporate staff, part-time employees, or consultants is acceptable.
  - It is desirable and expected, but not required, to use certified professionals such as certified safety professionals (CSP) or certified industrial hygienists (CIH) whenever possible when their expertise is needed.
- Funds
  - There should be a budget line item for safety and health expenditures for routinely used consumables, training, protective equipment, medical exams, and

the like in the work-site's budget. The budget should provide adequate funds to ensure all necessary safety and health expenditures are funded for the year, including a reserve for unexpected expenses. Based on the circumstances, there may need to be capital funds included to fund major safety and health enhancements such as engineering controls to mitigate a hazardous condition.

- Equipment
  - Management is also ultimately responsible to ensure that all recognized safety and health hazards in the workplace are being controlled. VPP work sites are expected to follow the standard hierarchy of controls (engineering controls, administrative controls, and personal protective equipment) to protect employees from the recognized hazards. As such, the VPP onsite evaluation team would expect to see appropriate hazard control equipment installed and in use at the work site. The equipment should be properly designed and installed, well maintained, and effective. Protective equipment should be readily accessible and available to employees at all times. Finally, employees should be trained on how to inspect, use, and maintain all hazard control equipment to protect their safety and health.

The site should have access to safety and health experts to assist in addressing specific issues. As mentioned previously, these experts are not required to be employees; they can be external resources such as corporate staff, safety and health consultants, insurance company professionals, and the like. The site must have a written plan on how and when these experts will be used. For example, the plan may state that an insurance company industrial hygienist will visit the site quarterly to collect personal noise samples from employees who work in noisy environments.

SAFETY AND HEALTH PLANNING

Safety and health, as a core value, must be incorporated in all aspects of the business. As such, safety and health must be integrated into the overall management planning for the facility. The employee responsible for safety and health at your facility should be invited to participate in the various planning activities for the site, such as:

- Setting production rates and goals
- Increasing or decreasing the workforce
- Introducing new equipment or products

The reason for this involvement is to ensure that all safety and health considerations are addressed at the proper time (the planning stage), when adjustments to ensure adequate protection for employees can be made easily, efficiently, and cost effectively.

Using setting production rates and goals as an example, the employee responsible for safety and health should be involved and ask the following questions:

- Are the production rates and goals reasonable?
  - Will additional equipment need to be added to handle the increased production rates?

  - Can the existing workforce handle the new rates and goals?
  - Will there be an increase in overtime hours worked (potential for additional injury/incidents due to fatigue)?
  - Will a new shift have to be added?
  - Will there need to be additional training for employees?
  - Do the new rates and goals increase ergonomic issues?
  - Is there a chance for additional employee exposure to physical or chemical stressors?
- Will operating procedures need to be changed to accommodate the new rates and goals?
  - Is there funding and time available to assess the hazards associated with the new procedures?
  - Is there funding and time available to revise the operating procedures?
  - Is there funding and time available to retrain employees?
- Will more employees need to be hired?
  - Is there funding to train and equip the new employees?
  - Is adequate time available to properly train the new employees?
- Will additional equipment need to be installed?
  - Is there funding and time available to do a preinstallation hazard assessment?
  - Is there funding and time available to do a job hazard analysis and write and train on operating procedures?
  - Will the installation of the new equipment pose any hazard, such as ergonomics?

ANNUAL SAFETY AND HEALTH PROGRAM EVALUATION

All VPP sites are required to conduct an annual self-evaluation. The self-evaluation is one of the most daunting tasks that must be completed for the VPP process. Many facilities find it very challenging to write a narrative, comprehensive evaluation that meets OSHA's requirements.

The purpose of the self-evaluation is to provide a thorough review of the effectiveness of the site's safety and health process over the past year. The reason OSHA requires the annual self-evaluation is to enable it to monitor the health of your safety and health process. The annual self-evaluation must be prepared and submitted by February 15 each year. OSHA will review the content of the annual evaluation to ensure that the VPP standards continue to be met and there are no significant systemic issues. The annual evaluation will also include the annual injury and illness rates from the OSHA 300 A log. OSHA will review the statistics to ensure the rates continue to fall below the BLS averages for your industry, and no upward trends are developing. If the OSHA VPP manager or VPP coordinator identifies any issues of concern, they may ask for some additional information or even visit the site to conduct a limited focus evaluation.

There is an additional self-evaluation requirement for any facility that is regulated by OSHA's Process Safety Management (PSM) standard, 29 CFR 1910.119. These facilities must complete the VPP PSM Program Supplement Self-Evaluation as well. The supplement requires that a number of questions, which are selected from OSHA's Primary PSM Inspection Priority List, concerning your PSM program be answered. These questions are very detailed and specific and require a substantial amount of time to be completed by a technical person familiar with your PSM program.

Most sites complete their annual evaluation at the end of the year; however, it can be done at any time of the year so long as one is completed every 12 months. The annual evaluation is completed in a narrative format, with the site thoroughly reviewing its entire safety and health process and identifying the strengths and weaknesses. To be effective and useful in improving your safety and health process, the annual evaluation should be done by a multidisciplinary team involving all levels of the organization and all departments. It is sometimes beneficial to invite an outside entity, such as the corporate safety and health staff or a consultant, to participate in the annual evaluation to provide an unbiased, independent viewpoint.

The team should divide responsibilities to complete the evaluation of the various elements of the site's safety and health process. As part of the annual evaluation, improvements should be noted as recommendations that can be evaluated to enhance the overall site safety and health process. The recommendations that are generated in the annual evaluation should be assigned to an individual for follow-up along with a targeted completion date. A person or group within the organization should be charged to follow up on the recommendations to ensure they are thoroughly evaluated and acted upon. A management representative should be assigned to monitor the progress in addressing the recommendations, ensuring employees are held accountable to complete their action items per the prescribed schedule.

The recommendations identified in the annual report can be used to form the basis of the safety and health goals for the following year. For example, if the annual evaluation identifies that additional job hazard analysis are required in the maintenance department, a goal can be assigned to an individual or team to complete a certain number of analysis in the upcoming year.

The annual evaluation should be made available to all employees for their review. The report can be made available by e-mailing it to employees, posting it on a bulletin board, or reviewing it at a safety meeting. Additionally, the recommendations should be updated periodically and also made available for employees to review so they can keep abreast of the various enhancements that are made to improve their safety and health.

Facilities that are applying for VPP recognition must attach the latest annual evaluation report to the VPP application. The annual evaluation that is submitted as part of the initial VPP application does not necessarily need to follow the prescribed format. However, it must be narrative in nature, provide a review of the entire safety and health management system, and include recommendations for improvement.

Additional information on the annual evaluation is included in Chapter 9.

EMPLOYEE INVOLVEMENT

One of the core requirements of the VPP is that employees must be involved in the safety and health process. Although this requirement may seem easy at first glance, it is actually one of the more difficult requirements to meet. The primary reason that this requirement is difficult is because the employee involvement must be active and meaningful. Their participation in the various components of the safety and health process is more than just coming to a meeting; they must fully participate and be engaged in the various safety and health activities.

Active employee involvement means that employees are involved in the ongoing safety and health activities. A system must be developed and implemented to allow for employees to become actively involved in the safety and health process. Management must make time available for employees to fully participate.

A well-developed safety and health process will have many components, and employees must be familiar with each of them. However, due to practical constraints, not all employees can be intimately involved with all aspects of each and every component of the safety and health process. To address this concern, an organization may create committees or teams to develop, implement, evaluate, and enhance the various elements of the safety and health process. Employees can then choose those activities they are most interested in and become involved in those committees or work teams.

Meaningful employee involvement means that employees have a say in the safety and health process. Meaningful employee involvement means that all employees, at least when addressing safety and health issues, are of equal status. For example, if a safety team was meeting to discuss a particular issue, and the team was composed of both managers and hourly workers, everyone would have an equal say in the proceedings, regardless of their rank or status in the organization.

Does the VPP force employees to actively participate in the various safety and health activities? Of course not! It would most likely be counterproductive if employees were made to participate in committees and teams when they had no desire or interest. The VPP expectation is that management provides several different opportunities for employees to be involved. The expectation is also that employees would take advantage of the opportunity and get involved. While it is not expected that all employees are actively and meaningfully involved in the safety and health process, a majority certainly should be. Lack of employee involvement when opportunities exist might be an indication of more significant labor management issues that the onsite VPP evaluation team would certainly investigate.

Why should an organization encourage employee involvement? There are many compelling reasons to involve employees in our safety process. Among these are:

1. *It can revitalize your safety process.* Even the best safety process will become stale over time. One of the easiest and best ways to help revitalize your safety process is to become a proponent of employee involvement. Now, instead of one or two "safety people" on staff, you have your whole facility on board helping to develop new ideas and concepts or enhancing existing ones.

2. *More help is always useful.* In today's workplace, everyone must do more with less. Many workplaces have employees performing two or even three jobs simultaneously. It is not very common to find a facility safety professional working on just safety. By encouraging employees to become involved in the safety process, we can increase the safety presence in each department on each shift, and multiply the amount of work that can be accomplished. These "safety ambassadors" can assist their fellow employees by using their knowledge and experience to make improvements in our safety and health process.
3. *Education.* Learning is a two-way street. On the management side, employees who become involved learn new skills that will benefit the company. The employees now become a safety presence that is thoroughly ingrained within the organizational structure 24 hours per day, 7 days per week. On the employee side, it allows their years of knowledge and experience to come to the forefront to help make improvements in the safety and health process. Employee involvement helps us transition from "managing for compliance" to "managing for safety" in that we become people focused rather than regulation focused. Employee involvement helps us to recognize the positives and correct unsafe acts through peer-to-peer interactions rather than with the stick management used in the traditional discipline-based safety system.
4. *Empowerment.* People want some control over their destiny. When employees have been empowered through involvement, they take ownership of the safety and health process and are much more likely to help drive continuous improvement. As employees come to realize that management truly cares about them and their safety and health, and values their thoughts and ideas, involvement will increase. We must explain very clearly what our expectations are, and what we would like to achieve through their involvement in the safety and health process. With empowerment comes responsibility, and with responsibility comes accountability. It is much easier to hold an employee accountable when they truly can accomplish a goal or project. Supervisors must also be held accountable to do what it takes to foster and maintain employee involvement among their subordinates.
5. *Perceptions are reality.* Show the employees management really cares. Employee involvement should not be sold as a tool for increasing productivity or sharing the workload among many people. The purpose is to improve our overall safety and health efforts. Remember the old adage about human nature: We react to the way others feel about us. If our employees perceive and believe management really cares about their safety and health, they will respond accordingly. If employees are not convinced management really cares, the truly meaningful employee involvement process is doomed to fail through half-hearted participation, commitment, and ownership by the employees. Do we really not want to see an employee injured, or are we really just trying to finish the year without an OSHA-recordable injury?
6. *Safety is ingrained.* If everyone at the facility has a role in the safety and health process, is empowered to make improvements, and is held accountable for his or

her safety performance, then safety has truly become ingrained throughout the operation and is not seen as a collateral duty or just another campaign or program to be implemented that will soon run its course.

7. *Safety becomes a core value.* We want to have safety become a core value, which is something that is so very important to us that we will never deviate from the value in spite of external influences. This is quite a shift from safety being a priority, which is something that can and will change when something more important comes around.
8. *Employee involvement fosters culture change.* As employee involvement becomes ingrained in the daily operations, safety and health becomes part of the daily routine both at work and at home. If we can demonstrate that safety and health really is an important core value within the organization, employees are more likely to understand the true value of working safely and no longer see safety and health as just another rule to be followed, as just another opportunity to be punished.

Although active and meaningful employee involvement may seem like the magic pill you may have been looking for to enhance your safety and health process, there are some potentially serious pitfalls that you must be aware of and be prepared to address should they arise.

1. *Ensure the involvement is meaningful.* No one wants to be a rubber stamp. No one wants to waste time working on something that never comes to fruition. Truly valuable employee involvement means that the participation must be "real" and the opportunity to implement new ideas and concepts must be present. Your organization must embrace the concept of thinking out of the box and must empower employees to have the authority necessary to accomplish their goals.
2. *Be ready for some new ideas.* Not all ideas and concepts that may be suggested are something you can use. There may be an idea that is so good, so cost effective, and will make such a significant impact that it can and will be implemented immediately. That situation is a no brainer; but, what do you do with that outlandish idea that is truly out in left field somewhere? You need to have a system in place to diplomatically address these off-base suggestions in such a way that you don't compromise the effectiveness and integrity of your overall employee involvement process. You must also be on the lookout for employees who may try to pursue other agendas under the guise of employee involvement.
3. *Make sure you have involvement opportunities.* Don't monopolize your involvement opportunities with just a few employees. It is very discouraging for an employee to volunteer to work on the safety and health process and not have the opportunity to get involved and present his or her great idea. Make sure your organization has enough opportunities for everyone who wants to be involved to get the chance to participate.

4. *Ensure your culture accepts empowerment.* When we allow people to get involved, we must be prepared to act on their suggestions for improvement. Traditional, centralized, formal organizational structures sometimes have difficulty in empowering employees to become involved, believing "that is not their job." Make sure your organizational culture will tolerate ideas from nontraditional sources to be fairly evaluated and implemented.
5. *Recognition.* Even more employee involvement will be generated when the successes are recognized and publicized for all to see. Everyone loves positive reenforcement, even if it is just a sincere thank you. Make sure you have a process to communicate your successes to all employees. Some sites dedicate a bulletin board in a well-traveled area to safety and health issues. A summary of the safety and health issue along with the solution suggested by an employee can be posted along with before-and-after photos of the area where the improvement was made as well as a photo of the employee who suggested the improvement. Make a concerted effort to publicize and recognize employees who are involved and contribute to the safety and health successes in public in front of their peers, perhaps at a safety meeting or a periodic "all-hands" meeting. Consideration should also be given to presenting a small token of appreciation and recognition, such as a free lunch or a company-shirt.
6. *Do not become compliance oriented.* Let the real value of employee involvement, which is peer-to-peer interaction based, shine. Leave the interpretation and implementation of compliance programs to the safety department. If there is a suggestion or question about a compliance issue, it should be forwarded to a subject matter expert to be addressed. Once the issue has been addressed, feedback should be provided to the employees.

For a safety and health process to be successful, all employees must take ownership. How do you get employees to take ownership of the safety and health process? Probably the best way is to get employees involved in all aspects of the safety and health process. Once employees tackle ownership and become involved, the next step is to empower them to act. Hourly employees will also need to be empowered to carry out their safety and health responsibilities. Empowerment, which is the authority needed to act, is another issue that must be addressed. Systems should be developed to provide a mechanism for employees to get involved in the safety and health process and be empowered to act.

This concept of involvement and empowerment may require a culture change in your organization. Some of the issues that may need to be addressed include:

- Will your organization accept hourly workers doing what has been traditionally management work, such as developing policies and procedures?
- Will your organization accept hourly employees having authority to enforce safety and health policies and procedures?
- Will your organization tolerate all employees, both management and hourly, having equal responsibility for safety and health?[1]

The VPP requirement is that all employees are active and meaningfully involved in the safety and health process in at least three ways. The definition of involvement is pretty liberal, with the exception of reporting hazards. Reporting hazards is not considered a means of employee involvement because it a right afforded them under the OSH Act.

Some examples of employee involvement opportunities include:

- Membership on safety and health committees or work teams
- Membership on the site's emergency response team
- Providing safety and health training to co-workers
- Participation on safety and health inspection teams
- Participation on job hazard analysis teams
- Writing or reviewing standard operating procedures
- Auditing contractors working onsite
- Participating on incident investigation committees
- Participating in the development of the site's annual evaluation
- Performing job safety observations of peers

In your VPP application, you will need to provide specific information about how employees are involved in the safety and health decision-making process at your site. Although it is clearly a management responsibility to set policy, if there is a decision to be made, employees should be consulted for their input. This concept is important to maintain employee commitment to the safety and health policy, and to demonstrate all employees are equal partners working together to ensure success. If employees are involved in the decision-making process, they will also be particularly helpful in resolving problems that may arise as they have a sense of ownership. Initially, this concept may only address safety and health issues, but over time can be extended to resolve problems in other areas of the business.

Employee participation is an essential and required element of participation in the VPP. Many of the participating companies have a very successful employee involvement process. With a successful involvement process, employees are no long seen as part of the problem but rather as part of the solution. Get your employees involved in an even more meaningful way today, and enjoy many more successes tomorrow!

SAFETY AND HEALTH COMMITTEE

Safety and health committees are not a VPP requirement for general industry sites. However, construction sites are required to have a safety and health committee. For the purposes of your VPP application, if there is a safety and health committee, you must give the date it was initially formed.

Safety and health committees can be an excellent tool to coordinate all of the safety and health activity at your site. It is helpful to develop a safety and health committee charter and mission statement, as well as long- and short-term goals and objectives. A well-designed safety and health committee will have a well-balanced membership, consisting of both management and hourly employees from all departments. Rank should be checked at the door. A manageable committee at a smaller work site should have a membership of no more than 10% of the workforce. Ideally, the chair of the committee should be an hourly employee. The membership should be carefully selected to avoid any perception that management is installing a "rubber stamp" committee. Potential members could volunteer and then be elected by the workers for a single fixed term of 2 or 3 years. The term limits provide opportunities for more employees to participate over time. Membership should be staggered so that continuity can be maintained.

A written procedure should be developed explaining the operation of the safety and health committee. A procedure will be needed delineating the number of members, how they will be elected, and how long they will serve.

A procedure will also need to be developed about the conduct of committee meetings, such as:

- The frequency of committee meetings. Most committees meet monthly.
- How many members must be in attendance to conduct business and vote on issues.
- Who will keep meeting minutes?
- Who is responsible for developing the meeting agenda?
- How will the minutes be made available for all employees to review, and who will be responsible?
- Who will be responsible for following up on the various committee action items?

As part of developing the safety and health committee's charter and long-term and short-term goals, thought should be given to the various roles the committee will have in regards to the site's safety and health process. Some common activities that safety and health committees are involved in include:

- Work-site inspections
- Auditing of employees and contractors
- Incident investigations
- Workplace hazard assessments
- Hazard correction
- Providing safety and health training

Once the committee's various activities have been identified, training of the committee members will be necessary. As a minimum, if committee members will be expected to conduct workplace inspections, hazard recognition training will be required. The training should be sufficient to provide committee members with an adequate amount of information so they can find unsafe conditions and OSHA compliance issues can be identified. Additionally, if committee members are expected to conduct incident investigations, training will be necessary. A record of all training completed by committee members should be maintained.

EMPLOYEE NOTIFICATION AND TRAINING

The VPP requires that management inform all employees and resident contractors that they are pursuing VPP recognition. The VPP also requires that employees have a minimum amount of training relative to the organization's involvement in the VPP:

- Information about the VPP, why the site is applying for VPP, and the benefits that accrue to the site and employees by participating in the VPP.
- Employees' rights and responsibilities under the OSH Act. Specifically, employees should be informed that even though VPP sites are removed from OSHA's programmed inspection list, employees still have the right to file a complaint with OSHA, and OSHA will conduct an inspection in the event of a fatality or "catastrophe."
- Employees' right to access OSHA standards, industrial hygiene sampling results, medical surveillance records, inspection results, and incident investigations.

Employee training on these three issues should be documented and available for review by the VPP onsite evaluation team.

CONTRACT WORKERS SAFETY

One of the precepts of the Voluntary Protection Programs is that an equal level of protection in terms of safety and health must be extended to contractors.

There are two types of contractors that are covered by the VPP. The first is resident contractors. Resident contractors are those contractors who are present at the work site on a regular basis. The two most common resident contractors found at VPP sites are janitorial workers and the security force. The second type of contractors found at VPP sites are the casual contractor. These contractors work on an irregular basis and are not present at the site on a regular basis. An example would be a contractor that comes to work one day per quarter to service your copying machine.

For the purposes of the VPP, contractors that work 1000 man-hours or more in any one calendar quarter are covered under the site's VPP application. You will need to list any contractor that exceeds this threshold in your application.

The pertinent issues relative to contractors include:

- Selection process
- Safety and health training
- Technical training

- Contractor auditing
- Annual performance review

The selection process for contract workers begins with the initial request for quotes for work. The VPP expectation is that there would be some mention of safety and health requirements and performance in the bid package. At a minimum, there should be wording that includes the following:

- The contractor will be selected by a number of criteria, including safety and health performance.
- The contractor will be expected to successfully complete a site-specific safety and health training program and have the necessary technical training and certifications for the work they will be performing.
- The contractor will follow all site-specific safety and health policies and procedures, as well as all federal, state, and local safety and health standards.
- The contractor will be periodically audited for safety and health compliance and performance; substandard performance may lead to removal from the site.

There must also be specific safety and health criteria for selecting contractors to work at the site. Each site must define what information will be reviewed, and what the disqualifiers will be. This determination should be based upon the potential hazards found at the host work site and the type of work being performed. The idea is to prequalify workers with an acceptable safety and health management system to work onsite.

Typical contractor selection criteria include:

- Review of the OSHA 300A log from the previous year. The contractor's TCIR and DART rates must be below the national average for his particular industry.
- Review of the contractors OSHA compliance history. All federal OSHA inspections for the most recent 5 years are listed in a database found at www.osha.gov. The site should review this database to determine if the contractor has received any OSHA citations.

  If the contractor has received an OSHA citation(s), he or she may not be immediately excluded from working at the site. Most VPP sites will accept a reasonable number of citations, depending on the type and severity. The general rule of thumb used at VPP sites is a company will be excluded from working at the site if the contractor received an upheld willful citation or he or she had a worker fatality directly related to the type of work to be performed at the site.
- The contractor should be required to submit an overview of his or her safety and health management system, including all policies, procedures, and the safety and health training program. The site should review the system to ensure the

contractor has an appropriate safety and health management system to protect workers from the typical hazards found in the workplaces.

Once the contractors' information has been received and reviewed, those meeting the site requirements should be placed on an "approved contractors list." The contractors on this list will be eligible to work at the site. The selection criteria should be reviewed annually to ensure any changes in the contractors' injury/illness rates, OSHA compliance history, and safety and health management system are reviewed to determine if they still meet the site's eligibility criteria.

Once the contractors have been awarded a contract to work onsite, they must successfully complete the site-specific safety and health training before being allowed to start work.

Typical training topics include:

- Site safety and health policy and contractor safety and health policy
- Entering and exiting the site
- Site hazard communication program
- Site emergency procedures, including:
  - Identification of the emergency alarm
  - Evacuation routes and assembly areas
  - Actions the contractor is expected to take in an emergency
  - Review of emergency equipment such as fire extinguishers, safety showers, and the like
- Review of relevant safety procedures, such as lockout/tagout, confined space entry, and so forth
- Incident reporting and investigation procedure

The training should be documented for review by the VPP onsite evaluation team. It is also a good practice to have the contractors successfully pass a written test at the completion of their training to verify they have understood the training.

Once the contractor has been trained and begins work, periodic safety and health audits should be conducted to ensure the contractor is in compliance with all of the various rules and regulations. The frequency of the audits should be determined by the site and will depend on the number of contractor workers on site, the duration of the job, and the hazards potentially faced by the workers.

The contractor audit team should consist of two or three employees, preferably one hourly and one management; this will facilitate meeting the employee involvement component of the VPP.

The contractor audit team should develop and use a checklist to ensure completeness and consistency between the various contractor audits. Any deficiencies found during the audit should be included on the checklist. The completed checklist should be forwarded to a responsible company official for follow-up. The site should develop a system for dealing with nonconformities that are identified during the audits. Ideally, deficiencies should be corrected immediately. For systemic issues, the contractor should be required to provide the site with an abatement plan. A responsible person should be delegated to ensure all corrective actions are addressed in a timely manner.

The site must include a contractor penalty provision in the contractor safety and health plan. VPP sites typically include either monetary penalties or a removal program for individual contractor employees or the company itself in the plan. Most VPP sites outline specific criteria for removing an individual contractor employee or the contractor company itself from the site for safety and health violations. Generally, VPP sites have a "three-strike rule" for minor violations (such as not wearing a hard hat) and an immediate removal rule for serious violations (such as not following the lockout/tagout procedure).

The contractor audit procedure should also have a requirement for an annual contractor safety and health performance appraisal. This procedure is a sunset review of the contractors' performance on site during the past year. This performance appraisal will be used in the selection process of the contractor for the following year.

The contractor program must also include:

- A process to ensure that all safety and health hazards created by the contractor are promptly identified, controlled, and corrected.
- A method to ensure that all contractor injuries, illnesses, and other incidents are promptly reported to the site and investigated.[2]

As part of your VPP application, you are required to submit a list of "applicable contractor companies" (those that work more than 1000 hours in a quarter) and the approximate number of contractor employees onsite at the time of the application or the most recent calendar year.

The VPP onsite evaluation team will certainly review contractor work activities being conducted onsite. Besides a review of the contractor selection and training documentation, the onsite evaluation team will also review the contractors work site and interview contractor workers to ensure they are knowledgeable about the VPP and are in compliance with the various rules and regulations.

As a word of caution, do not encourage contractors to stay home during the VPP onsite evaluation. If the onsite evaluation team determines that contractors are normally present, and they are not present during the evaluation, the team is likely to return at a later date to review the contractors. Additionally, this type of behavior would be contrary to the philosophy of VPP and would likely lead to an unfavorable opinion of the company, which could influence your level of participation in the VPP.

REFERENCES

1. B. T. Bennett. Employee Involvement. *The Leader*, **12**(3): 40–41 (2003).
2. B. T. Bennett. Orientation Program for Casual Contractors in the Chemical Industry. *Professional Safety*, **45**(8): 24–28 (2000).

Exhibit 4.1 Sample Safety and Health Policy

As a core value, the safety, health, and well-being of our employees and clients is of paramount importance.

PRINCIPLE 1 – We conduct operations in a manner that protects the safety, health and well-being of our employees and our customers

- Our top priority is the safety, health, and well-being of our employees and clients.
- State of the art safety and health programs and systems will be implemented to protect our employees. Mere compliance with applicable laws and regulations is not enough.
- Safety and occupational health will be integrated into all our work activities.
- Without fear of reprisal and discrimination, employees shall have the authority to stop work activities that would expose them to imminently dangerous hazards.
- Emergency situations in the office and in the field will be handled appropriately.
- Our goal is to eliminate occupational injuries and occupational illnesses for our employees and protect them when elimination is impossible.

PRINCIPLE 2 – Everyone is committed to the process

- Management will be accountable for safety and occupational health through demonstrated commitment to the safety and health management system.
- All employees will recognize that working safely is a condition of employment, and that they are accountable for their own safety and the safety of those around them.
- Resources will be available to achieve our goals.
- Potential risks in all of our operations (including inspections) will be addressed.
- All employees will be trained to perform their jobs safely.
- Off-the-job safety will be promoted to extend and reinforce safety and health consciousness.

PRINCIPLE 3 – Performance is measured and the results will be used toward on-going improvement of our Safety and Health Management System

- Effective performance measures will be utilized toward improving our safety and health management system.
- Behaviors, work processes, and management systems will be routinely examined.
- All incidents and near misses/hits will be investigated to determine contributing factors and to improve ongoing prevention efforts.
- Open communication regarding safety and health performance will be fostered at all levels.

Exhibit 4.2 Sample Long-Term Organizational Goal with Supporting Short-Term Departmental Goals

Long term organization goal:

- Develop a preventative maintenance program

Short term department goals:

- Assemble a preventative maintenance program committee
- Develop a preventative maintenance program charter
- Provide training to the preventative maintenance program committee
- Develop preventative maintenance standards and testing methods
- Purchase required preventative maintenance program testing equipment
- Train preventative maintenance program inspectors

Exhibit 4.3 Employee Safety and Health Performance Appraisal Report

Employee :________________________
Date :______________________________
Performance Period :__________ to __________

O = Outstanding
E = Exceeds Expectations
M = Meets Expectations
I = Needs Improvement
N/A = Not Applicable

		O	E	M	I	N/A
A	**REGULATORY DUTIES**	**O**	**E**	**M**	**I**	**N/A**
1	Ensured operations are in compliance with company, federal, state and local health and safety laws, regulations, and standards through scheduled inspections, good housekeeping, and accurate record keeping.					
2	Responded immediately to all incidents; implemented corrective actions; investigated all incidents that occurred in their area promptly and thoroughly; and submitted written reports in a complete and timely manner.					
3	Ensured a safe work environment by timely follow-up of all safety, health, and environmental deficiencies in their area, and by ensuring that all safety, health, and environmental rules and procedures were followed.					
4	Ensured that all OSHA hazard communication labeling was in place on all containers, raw materials, finished goods, process samples, and pipes and that all process sample points were labeled as required.					
5	Completed assigned safety action items in a timely manner.					
B	**SPECIFIC DUTIES**	**O**	**E**	**M**	**I**	**N/A**
1	Implemented new or revised safety and health rules and/or procedures within the prescribed time limits.					
2	Ensured all work permits have been completed properly; that employees have prepared jobs properly; have worn all prescribed PPE; completed work within established procedures; and the work area has been returned to a safe condition.					
3	How is housekeeping in the employee's area?					
4	Measure the employee's participation in safety and health incident investigations by considering the thoroughness of investigations, quality of the investigation, and the implementation of corrective actions.					
5	Monitored subordinate employees' safety and health performance and held them accountable for their performance.					
6	Participated in committees; attended safety and health meetings; communicated new and revised rules and procedures to subordinate employees.					
C	**ATTITUDE**	**O**	**E**	**M**	**I**	**N/A**
1	Promoted safety and health consciousness through appropriate attitude and example setting.					
2	Prevented safety and health violations by positive reinforcement of good performance, observed actions of employees, and took a leadership role in corrective instruction.					
3	Responded to employee safety and health concerns and demonstrated commitment to employee welfare. Achieved and maintained an atmosphere of heightened awareness and cooperation through interpersonal leadership skills.					
4	Ensured safety and health policies and procedures were understood and complied with by the employee and their subordinates.					
5	Ensured a positive attitude existed towards acceptance of and response to identified deficiencies noted on inspections, audits, etc. by the employee and their subordinates.					
D	**TRAINING DUTIES**	**O**	**E**	**M**	**I**	**N/A**
1	Conducts safety meetings and tool box talks as required.					
2	Completed all of their individual annual required regulatory refresher training within the prescribed time limits.					
3	If in a supervisory position, completed required regulatory training within the prescribed time limits for all subordinate employees.					
E	**GOALS**	**O**	**E**	**M**	**I**	**N/A**
1						
2						
3						
	OVERALL RATING					

	Employee:
	Supervisor:
	Safety and Health Appraiser:

Next Year's Goals			

5

WORK-SITE ANALYSIS

Before any hazards can be eliminated or controlled they must be identified and evaluated. There are several methods used to identify workplace hazards, and there are several intervals or steps that may be followed. In terms of cost-effectiveness and efficiency the best time to start the hazard analysis process is the design phase. When designing a new factory or new equipment, a design review team should be created. The design team should include both hourly and management employees, as well as safety and health experts, to ensure that there is nothing in the design that will introduce a new hazard into the workplace. It is especially critical to have employees that will be working on or using the new equipment participate, as they will bring a unique perspective to the analysis. Once the factory or equipment is ready to startup, the process should be reevaluated for hazards that were created when the equipment was built and installed, and may not have been identified in the initial hazard analysis.

A baseline survey and analysis is a first attempt at understanding the hazards at a work site. It establishes initial levels of exposure (baselines) for comparison to future levels, so that changes can be recognized. The hazards identified during the initial hazard assessment may be physical, environmental, or chemical. A comprehensive review of all workplace operations must be undertaken to identify all potential employee exposures to the identified hazards. It is important that these hazard assessments be performed by trained, experienced, and competent individuals. When performed, all buildings, rooms, processes, machines, tools, chemicals, products, and employee activities must be reviewed. Based on the results of the initial hazard assessments, additional more detailed assessments must be performed. There are several

Preparing for OSHA's Voluntary Protection Programs. By Brian T. Bennett and Norman R. Deitch
Copyright © 2010 John Wiley & Sons, Inc.

hazard assessment techniques available to identify workplace hazards. They include qualitative hazard analyses, job, task, process, or phase hazard analyses, industrial hygiene sampling, and occupational health assessments.

In addition to these hazard assessments, much can be learned about the hazards in the workplace by examining the safety and health history of the workplace. There are several learning opportunities available in most workplaces:

- Incidents that have occurred can be reviewed to try to determine what the causes were so that corrective actions can be implemented to prevent recurrences of similar incidents.
- Industry trends can be reviewed.
- Employees can provide significant input into the hazards associated with their jobs and tasks.
- Outside experts can be consulted for their insights.
- Government agencies can provide valuable resources.

The worksite analysis element of the OSHA VPP contains seven separate subelements that must be in place for participation. They are:

- Baseline hazard analysis
- Hazard reviews of routine jobs, tasks, and processes
- Hazard analysis of significant changes
- Routine self-inspections of the entire workplace
- Procedures for employees to report unsafe conditions
- Investigations of all incidents
- Trend analysis of all safety and health indicators

BASELINE HAZARD ANALYSIS

The baseline hazard analysis should be performed upon the startup of a plant and upon the introduction of significant changes to the workplace, including new processes or equipment. This may also include the expansion of the workplace, significant revisions of the equipment, installation of new equipment, introduction of new chemicals, or changes to the process. Each of these changes may eliminate some older hazards and may actually be intended to do just that. However, it is also possible that they will introduce new hazards that then have to be evaluated and controlled. On the face of it the workplace may seem to be a safer one, but in actuality the modification may simply change one hazard for another one.

Although the VPP and the OSHA Safety and Health Management System Guidelines both encourage the use of employees to perform such hazard analyses as inspections and job hazard analyses, it must be stressed that these are technical functions, and those that are performing them must be trained in the process.

The analysis should start with the qualitative hazard analysis, which is a comprehensive review of facility operations to identify all possible employee exposures to unsafe equipment, chemicals, or environments; both health and safety issues should be identified. It is important that this qualitative hazard assessment (QHA) is performed by a trained, experienced, and competent safety and health professional as a member of the team. When performed, all buildings, rooms, processes, machines, tools, products, and employee activities should be reviewed. Based on the results of the QHA, where employee exposures to unsafe or unhealthful conditions are determined to be possible, a sampling plan will need to be developed specifying the necessary quantitative exposure assessments (industrial hygiene sampling) or detailed job hazard analyses that will be performed in order to identify actual exposures.

The QHA is intended to identify all hazards that may be present in the workplace and lead to more detailed analyses. When performing the QHA, care must be taken to properly organize the effort. The first thing to consider is who is going to perform the QHA. In view of the critical and technical nature of the process, it is recommended that the job be assigned to a safety and or health professional that is familiar with the industry, the equipment, or the processes in use at the workplace. That person should be supported with a team of hourly and management employees that are familiar with the process. The team should consist of the local safety manager, operators, and maintenance staff. It may also be helpful to include outside resources such as confined-space rescue services and hazardous materials response organizations.

The QHA team will devote their attention to several factors:

- Physical layout—this will include:
  - Consideration for emergency egress
  - Lighting
  - Ease of navigation around the equipment
  - Smoothness of the flow of work
- Operating equipment—this will include:
  - Identifying pinch points and other point of operation exposures
  - Effectiveness of guards
  - Use of electronic or electric interlocks
  - Location of emergency stop buttons or cords
  - Electrical connections
  - Location of electrical service disconnects
  - Identification of all energy sources
  - Identification of all lockout points
  - Location and access to maintenance points
- Occupational health considerations—this will include:
  - Review of material safety data sheets (MSDS) for all chemicals used or stored in the workplace

- Review of all processes to determine potential exposure to harmful chemicals caused by the process, such as:
  - Hexavalent chromium as a result of welding stainless steel
  - Silica dust as a result of mixing or dry cutting concrete
  - Carbon monoxide from internal combustion engines such as powered industrial trucks (fork trucks)
- Noise surveys to determine potential exposures to excessively high noise levels
- Design of workstations regarding the potential exposure to unsafe body mechanics, such as:
  - Lifting and twisting under load at conveyor tables
  - Lifting heavy materials
  - Standing stationary for extended periods
  - Using a phone while keyboarding as a receptionist may have to

- Environmental considerations—this will include:
  - Exposure to temperature extremes
  - Exposure to humidity extremes

Each of these factors must be examined and inspected in detail. The observations and reviews will be documented so that preventive measures can be planned and implemented. Some of those preventive measures would include designing and installing point of operation guards and barriers, relocating equipment to provide better means of egress and reduce noise levels, improving lighting, improving ventilation, changing from one chemical to a less harmful one, redesigning workstations, developing an industrial hygiene sampling plan for a more detailed review of potential harmful chemical and noise exposures, installing clean rooms to avoid contamination of food and beverages, and other similar actions.

Based on the qualitative hazard assessment, additional detailed full-shift surveys should be planned and performed. These surveys must not be limited only to industrial hygiene activities. The surveys should also address safety-related concerns raised during the qualitative hazard assessment. A formal industrial hygiene sampling plan for the completion of a quantitative hazard assessment should be developed based on the findings of the initial air sampling and noise area monitoring tests to ensure that employees are not exposed to air contaminants and noise levels that exceed OSHA's permissible exposure limits. The sampling plan should be based on a rationale that identifies those potential exposures that pose the greatest risk. That consideration must take into account the actual potential for exposure given the engineering controls in place, the frequency of the off-ending process, and the number of employees exposed. The sampling plan should also identify the jobs and tasks that will be sampled and the schedule for the sampling. The schedule must consider the time of day, the day of the week, the time of year, and the ambient weather conditions. The idea is to obtain exposure information during the periods of low and high exposures.

For example, the sample results for carbon monoxide in an open garage in the northern area of the country during the summer may be relatively low, but the results will most likely be higher in the winter when the doors are closed. Another example is the potential variances in sample results when testing performed at the beginning of a work day as compared to the end of the work day when air concentrations may be higher because of the cumulative effect.

It is important that industrial hygiene sampling results be compared to up-to-date health indicies. In the spirit of the VPP, that means that sampling results should be compared against recommended exposure limits such as the threshold limit values (TLVs) published by the American Conference of Governmental Industrial Hygienists (ACGIH). The reason that VPP prefers comparison to the TLVs is that they are updated every year based on the latest scientific data. The OSHA permissible exposure limits (PELs) found in 29 CFR 1910 are based on health data that may be more than 40 years old and are reflective of the current knowledge concerning health effects of various stressors.

If sampling results indicate overexposures (or levels even approaching exposure limits) of the more restrictive of the measurement criteria, corrective actions will need to be implemented and documented. These recommendations may take the form of engineering controls, administrative controls, work practice procedures and programs, and/or PPE. These controls, and others, are discussed in greater detail in Chapter 6.

All sampling must conform to nationally recognized procedures and techniques, and all samples must be analyzed by a recognized and accredited testing laboratory. Companies can also do their own sample analyses if their labs have been certified to meet the national standards. All sampling and similar work-site analysis activities should be documented as a form of verification that the work was performed and to provide an audit trail for reviewers.

Sampling data must be reviewed with all affected workers. That includes all employees that were sampled, as well as employees that perform similar jobs. It is best if all personal information is removed from the sampling data report; then the results are shared with the affected employees. There should be documentation that the results were not only shared with, but explained to, all affected employees.

Development of the sampling plan should also consider employee input. Employees can provide a great deal of insight to the process of developing the sampling plan. Employee interviews can use the following suggested questions to obtain valuable information:

1. Do you come into contact with any potentially dangerous chemicals, substances, or harmful physical agents such as dust, radiation, or noise?
2. If so, what are they?
3. Were they ever quantified by sampling for exposures?
4. How often does the exposure occur?
5. When is the greatest exposure or potential for exposure?

6. Where is the greatest exposure or potential exposure?
7. Do you feel that management has provided enough protection for you?
8. Have you ever raised concerns about it?
9. Have you ever seen industrial hygiene surveying or monitoring being done in your workplace?
10. Was it just once or is it routine?
11. If just once, was it in response to a specific problem?
12. If a specific problem, what was it?
13. If routine, how often?

The baseline safety surveys must consider the potential exposure to moving equipment and machinery, the use of power tools and equipment, material handling equipment such as fork trucks, cranes and hoists, electrical equipment, maintenance of equipment, tool and machine adjustment, potential for falls from one elevation to another, ladder use, and all other physical hazards.

The hazard assessment process is not complete once the baseline assessment is complete. Quite often, conditions and employee activities change over time as the equipment is run. Therefore, a routine and periodic assessment should be performed to ensure that the original (baseline) assumptions are still true, and there are no hazards being posed to employees. Make sure that all jobs, even those in the office area that are not considered high risk, are periodically reviewed.

ROUTINE HAZARD ANALYSIS OF JOBS AND TASKS

It is very common and relativly easy to recognize the serious hazards in the routine jobs and tasks in a workplace. With very little training most managers and workers would be able to recognize the obvious point of operation pinch points and the potential for falls at open-sided floors. All jobs and tasks have hazards associated with them, even those considered the most mundane. Unfortunately, most workers and managers either do not recognize the more routine but nonetheless serious exposures as hazards or they do not consider them as serious enough to merit their attention. Consider the following and think about your experiences:

- The office manager incurring an amputation while using a manual paper cutter
- The receptionist who has a repetitive neck injury from cradling the phone between her ear and shoulder
- The shipping clerk who pulls her back lifting several heavy boxes
- The manager who receives an electric shock when spilling the water as he is changing the bottle on the water cooler

These and other more serious jobs and tasks must be evaluated to identify the hazards associated with them to be able to develop and implement the appropriate controls to protect employees from exposure to those hazards. There are several techniques and tools used to identify those hazards. Before the techniques and tools can be used, a system should be developed to indentify each job and task and assign a risk factor to that task. A useful process to start with is an inventory of all jobs and tasks performed by each department.

Clarification of the distinction between a job and a task must be understood. The easiest way to distinguish between the two would be to think of the job as a group of tasks. For example, the description of the job of a fork truck operator is to operate the truck to move material. That job consists of several tasks including:

- Inspecting the truck before beginning to operate it
- Changing a fuel tank or disconnecting the battery charging cable
- Driving to the location of the material
- Picking up the material
- Driving the material to its destination
- Unloading the material
- Moving on to the next load

Each of these tasks are directly associated with the job of the fork truck operator and have numerous steps that must be performed. Each of these steps should be reviewed to identify the associated risks and specify how the employees shall be protected. Although it is not necessary to separate the individual tasks from the job, the more complex jobs are easier to evaluate when broken down to the more specific tasks.

Once an inventory of all jobs is completed, it should be further broken down to the specific tasks. Upon completion of the detailed inventory, those tasks that are more likely to cause injury or illness must be identified to be able to schedule the hazard analysis of tasks based on a priority of addressing the most serious hazards first.

The priorities for determining the schedule of hazard analysis include:

- *Historical* Determine the history of risk and adverse consequence for this piece of equipment or process. You should consider not only the site's experience but also the corporation's as well as the industry's experience.
- *Perceived Risks* Sometimes a first blush is right on target. If the employees or the hazard analysis team feel this equipment or process has some inherent risks, they should be evaluated early in the process.
- *Complexity of the Task* Typically, the more complex the task in terms of the number of steps, the difficulty in completing the steps, or the risk associated with each step is an indicator of the amount of risk present in the job.
- *Frequency* If the task is considered nonroutine, it is not performed on a regular basis. While each facility defines the time period that constitutes a nonroutine task, the standard definition is conducting the task less than once or twice per year. The concept with nonroutine tasks is that since employees do not conduct them on a regular basis, they are inexperienced at performing

them. Since they are inexperienced, they will likely need additional refresher training, including a review of the safety and health risks, immediately before performing them.

- *Regulatory Requirement* Some OSHA standards require routine hazard analysis. One example is the Process Safety Management Standard (1910.119), which requires that a process hazard analysis be performed for each process at least once every 5 years. That standard also requires that operational employees participate in the process hazard analysis.

Each task should be measured for its degree of hazard to determine the prioritization of the analyses by task. To determine the degree of hazard of each task, the following must be given consideration:

- Review the OSHA logs, first-aid logs, accident reports, near-miss reports, and other similar logs and reports:
  - Identify those tasks that have resulted in the greatest number of injuries or illnesses.
  - Identify those tasks that have resulted in the most severe injuries or illnesses.
- Review industry reports of injuries and illnesses.
- Review the OSHA reports of injuries by standard and industry.
- Survey the Employees.
- Review the MSDSs for all chemicals used in the workplace.

Once the degree of hazard of each task has been determined, list them in the order of priority. Address first those that are the most likely to cause serious injury or illness and work down to those tasks that are either less likely to cause injury or if an injury is likely to occur it will be minor, requiring nothing more than some first aid and minimal follow-up.

The job hazard analysis (JHA), while not a very technical process, is nonetheless a very detailed one. There are important considerations and evaluations that must be made, and those that are completing a JHA should be trained on the process. They must be familiar not only with the job or task being reviewed but the forms and terminology that are used for a JHA. For example, a typical JHA has at least three columns:

- *Step* The step should be described as an action that is carried out. Many times it is incorrectly described as the hazard.
- *Hazard* The hazard should be a description of what may happen if the step leads to an injury or an illness that may result from the step if no protective measure is taken. It usually contains a term that describes a medical condition such as a hearing loss, a strain, or a laceration.
- *Protection* Last section is a description of how the employee should perform that step or what PPE should be used to avoid the hazard described.

Once the JHA is completed, the employees performing the task should be trained on the JHA so that they may be in a better position to protect themselves. The JHA could also be incorporated into a standard operating procedure to provide more detail not only about the procedure for the work to be performed but the safe way of performing it. The JHAs should be reviewed on a routine basis to ensure they still adequately describe the steps of the task. The review should also consider new methods of protection against the hazards. There is no specific criteria for the frequency of review for JHAs. At a minimum, they should be reviewed whenever there is a change in the process, equipment, or chemicals in use. The JHA should also be reviewed whenever there has been an incident related to the task it covers. The incident investigation should also include a review of all applicable procedures, including the associated JHA. Without any specific cause, such as an incident or change of the task, the JHAs should be reviewed no less frequently that every 2 years. The frequency would depend on the number of JHAs in use and the hazardousness of the task or process. There are some OSHA standards that require a review of the hazard analysis with a specific frequency. The OSHA Bloodborne Pathogen Standard (29 CFR 1910.1030) requires that a review of new needle technology be performed regularly to identify better protections against needle sticks, and the Process Safety Management Standard (29 CFR 1910.119) requires that process hazard analyses be reviewed at least once every 5 years.

HAZARD ANALYSIS OF SIGNIFICANT CHANGES

Most workplaces undergo changes that involve either adding new product lines, replacing equipment with newer technology, replacing the chemicals or process used, or expanding or revising the physical plant. OSHA also considers tasks that are performed less than once per year as a nonroutine and falling into this category. These changes must be evaluated to identify any new hazards that they may introduce to the workplace. These analyses, like those for routine tasks, should identify the proper control methods for the identified hazards.

The level of detail of the analysis should be commensurate with the perceived risk and number of employees affected. This practice should be integrated in the procurement/design phase to maximize the opportunity for proactive hazard controls. The VPP process requires that the applicant explain how new or significantly modified equipment, materials, processes, and facilities are analyzed for hazards prior to purchase, installation, and use. One example of the importance of this type of review is illustrated by the following true situation:

> A manufacturing plant replaced an older piece of equipment with a newer state of the art machine with well-designed safety controls, including guards and interlocks. However, no analysis was performed to address the actual placement of the machine. The operational and maintenance employees were not involved in the process. A problem arose when the maintenance mechanics found that their access to the maintenance points was limited, causing an unsafe condition. The company had in effect improved the safety of the machine operators at the risk of the maintenance crew. The end result was that the equipment had to be relocated to allow the mechanics safe access to the equipment at an additional substantial expense. This could have been avoided if the company had performed an initial hazard analysis during the design and layout

phase of the equipment, including all those that would be involved in the use of the machine.

The review for hazards associated with significant changes is not specifically required by most OSHA standards. However, the Process Safety Management Standard does require a process called Management of Change as follows:

29 CFR 1910.119 (l) **Management of Change.**

(1) The employer shall establish and implement written procedures to manage changes (except for "replacements in kind") to process chemicals, technology, equipment, and procedures; and changes to facilities that affect a covered process.

(2) The procedures shall assure that the following considerations are addressed prior to any change:

(i) *The technical basis for the proposed change;*

(ii) *Impact of change on safety and health;*

(iii) *Modifications to operating procedures;*

(iv) *Necessary time period for the change; and*

(v) *Authorization requirements for the proposed change.*

(3) Employees involved in operating a process and maintenance and contract employees whose job tasks will be affected by a change in the process shall be informed of, and trained in, the change prior to start-up of the process or affected part of the process.

(4) If a change covered by this paragraph results in a change in the process safety information required by paragraph (d) of this section, such information shall be updated accordingly.

(5) If a change covered by this paragraph results in a change in the operating procedures or practices required by paragraph (f) of this section, such procedures or practices shall be updated accordingly.

TASK HAZARD ANALYSIS

The typical method to identify the hazards or risks of each task and develop methods of protection is referred to as a job or task hazard analysis. Certain industries use similar procedures but refer to the process with unique terminology. For example:

- *Construction Industry* Phase hazard analysis is focused on the constantly changing workplace and activities of a construction project.
- *Chemical Industry* Process hazard analysis is focused on a comprehensive evaluation of the entire process that involves the use of specific chemicals that are considered to be hazardous. OSHA has also mandated process hazard analysis be completed for the extremely hazardous chemicals identified in OSHA standard 29 CFR 1910.119, Appendix A, List of Highly Hazardous Chemical,

Toxics and Reactives. The list identifies the threshold quantity (TQ) for each of the chemicals that triggers compliance with the standard.

Regardless of the title of the process used, the techniques are very similar. Employee involvement in the hazard analysis process is extremely important. Not only does it provide employees an additional means to be actively involved in safety and health, the employees are usually in the best position to describe the operation or process under review. Another important consideration is that at least one person involved in the process must be trained to be able to recognize hazards and in the task hazard analysis process.

The first step in performing the task hazard analysis is making a general assessment of the working conditions. This includes observations of:

- Housekeeping
- Adequate lighting
- General condition of equipment
- Employee knowledge of the job
- Use of the required PPE

Having the employees perform the task under review:

- Using action words, describe each individual task or steps as they are performed:
  - Lift a 50-pound box from a pallet on the floor.
  - Place the box on the work table to the side.
  - Open the box with a box cutter.
  - Remove the material and place it on the table.
- Examine each step to determine the possible resulting consequences using injury/illness descriptions:
  - Back strain from lifting a heavy load.
  - Strain from twisting under load.
  - Cuts from the exposed blade of the box cutter.
  - Arm strain from lifting the material out of the box.
- Using the hierarchy of controls, list for each step the safe procedures or PPE to protect employees from the identified hazard:
  - Place pallets on self-leveling platforms to maintain a proper height of the material.
  - Relocate the work table to the front to eliminate any twisting.
  - Use self-retracting box cutters and Kevlar® to prevent cuts.
  - Use a tilting table or a vacuum lift to eliminate the stretching motion.

Once the task hazard analysis has been completed, all employees must be trained on the procedures and the use of any new equipment, tools, or personal protective equipment. Task hazard analyses should be reviewed on a routine basis such as annually or every 2 years to ensure that there have been no changes that can lead to additional employee exposures. Task hazard analyses should also be reviewed anytime there has been an incident involving the task covered by the task hazard analysis.

ROUTINE SELF-INSPECTIONS OF THE ENTIRE WORKPLACE

The Voluntary Protection Programs require that general industry workplaces undergo some type of inspection activity at least monthly, with the entire workplace inspected at least quarterly. Construction projects, as a result of their regularly changing conditions, must undergo some inspection activity at least weekly with the entire project inspected at least monthly. Routine inspections provide excellent opportunities for hourly employees to become more actively involved in the safety and health activities. They are also ideal resources because of their regular presence on the work floor. Supervisors and managers should also participate in inspections to demonstrate their personal commitment to the safety and health management system.

There are three major types of safety inspections: pre-use inspections, preventive maintenance (PM) inspections, and comprehensive (facility walk-around) inspections. Preventive maintenance inspections are considered a safety and health control measure and will be discussed in Chapter 6.

Pre-use inspections are those required to be performed by the operator or user of certain types of equipment prior to each use (or at least the first use on each shift). Examples of equipment requiring pre-use inspections include personal protective equipment, ladders, vehicles, interlocks, and lifting equipment. With powered industrial trucks being the one exception, pre-use inspections do not need to be documented, but there must be some method to verify that these inspections have occurred. This is often confirmed through employee interviews. It is suggested that the site ensure that it has identified all equipment subject to pre-use inspections and that it remind all affected employees of the requirement to perform pre-use inspections. It is also important for all affected persons to be trained in how to properly perform the pre-use inspection and how to identify acceptable versus unacceptable inspection conditions. Pre-use inspection training for affected persons must be documented.

The third type of inspection, the facility walk-around, is the one usually thought of when discussing inspections. It is some times referred to as a walk-down or housekeeping inspection. It may also be called an audit, which is a term that has become almost synonymous with inspections. These inspections include observations of the workplace, operating equipment, emergency equipment, employee behaviors, and any other elements that may have an effect of the safety and health of the workers. The following discussions will be related to this type of inspection.

It has been found that those inspecting the same area all of the time tend to overlook unsafe conditions. That is not a result of their lack of knowledge or training but more likely a result of their familiarity with the hazard. They have become so accustomed to seeing the same thing, they don't recognize the reality of the hazard. An example may be a pipe that runs across a walkway and poses a tripping hazard. Everyone in the area knows that it is there, and they just have to step over it. Obviously, the better solution would be to relocate the pipe or the path of travel, or to install a step over the pipe. Therefore, it is strongly recommended that those that are performing the inspection do so at areas of the workplace that they may not be that familiar with. This is the principle that a "new set of eyes" will be able to see the "trees through the forest." Probably the last person that should be solely responsible for performing routine workplace inspections is the safety and health manager. Although inspecting the work site for unsafe conditions is a key responsibility for the safety and health manager, he or she should not be the only person conducting these inspections. It is more effective for the manager to ensure that all levels of employee, from manager to line employee, conduct work-site inspections. The more appropriate function for the safety and health manager is to be responsible for developing and ensuring the effectiveness of the inspection program. If it works effectively, there will be a lot more individuals looking for and observing hazards. That is not to say that the safety and health manager should not perform any inspections. He or she must make some inspections to be able to properly gauge the effectiveness of the program.

All employees expected to perform inspections must be trained on the inspection procedures, checklists, and forms. They should also be provided with some level of formal hazard recognition training. These trained employees will tend to become better safety and health advocates and will add to the effectiveness of the inspection process.

There should be some form of documentation completed to memorialize the inspection. Unless the inspection activity is documented, OSHA will take the position that the VPP inspection requirements were not met. It is also critical that all inspection findings are tracked until the corrective action has been completed. The documentation and tracking may be done either with very technical software or with a very simple system of maintaining the inspection reports and checking off the findings as they are corrected. Although some facilities use checklists to document the performance of inspections, consideration should be given to limiting their use. The downside of using checklists is that employees conducting the inspection tend to focus only on the items listed on the checklist and routinely miss other potential hazards that are not on the checklist.

All findings of the inspection must be documented, including those that are corrected immediately. The reason for that is to ensure that the hazard has been eliminated and to help develop a larger pool of information to use in the safety and health trend analysis. The documentation should identify the offending condition, the person responsible for correcting it, the recommended corrective action, and the assigned completion date.

If an unsafe condition is found during an inspection and can cause injury to an employee, safeguards must be put in place immediately. For example, if a walking surface is damaged and an employee can step in it and twist an ankle, the area should be barricaded immediately. The barricade provides immediate protection; entering a work order will result in the area being repaired permanently.

The OSHA VPP onsite evaluation team will spend a significant amount of time evaluating the action item tracking system. The main concern, in regards to inspections, is that the unsafe conditions that have been found are repaired in a timely manner. Although there is no "magic number" as to how many hours or days are appropriate, the general rule is the more serious the hazard the quicker it is repaired.

There must also be a system in place to inform employees of the hazards discovered during the inspection so they are aware of them and can take the appropriate precautions to avoid injury. This awareness of findings can be accomplished in many ways, including posting the information on bulletin boards, making an entry into log books, and covered in a safety meeting or toolbox talk.

An example of a hazard finding and correction for a general industry facility may be:

- *Finding* Numerous electrical service disconnects not adequately labeled to identify the voltage and purpose of the device
- *Recommended Action* Label all deficient service disconnects
- *Responsible Person* Maintenance manager
- *Assigned Completion Date* Two weeks from observation

An example of a hazard finding and correction for the construction industry may be:

- *Finding* Lack of ground fault circuit interrupter (GFCI) for carpenters' portable power tools
- *Recommended Action* Use of GFCI-equipped extension cords
- *Responsible Person* Carpenter supervisor
- *Assigned Completion Date* Two hours from observation

PROCEDURES FOR EMPLOYEES TO REPORT UNSAFE CONDITIONS

All employees are provided the right to report unsafe conditions by the OSH Act. The VPP expects that the workplace will have an effective system in place to enable the employees to do so. They system may be either formal or informal, but should meet the following criteria:

- The employees are aware of it.
- They confirm it works effectively.
- There is no fear of reprisal.
- The employees know how to recognize a hazard.
- Hazards are corrected promptly.
- Interim protections are provided until final correction.
- There are rewards for suggestions.
- The system can be used anonymously.
- Feedback is given to the employee reporting the hazard.

Unless the employees are aware of the reporting system, it will not work. The employees should be trained, with documentation, on the reporting system when they begin work and annually thereafter. The system may range from the very simple to the very formal. Some smaller workplaces simply have the employees notify a supervisor or maintenance department about the hazard. Others may use a form that is placed in a suggestion box that is accessed by the safety and health manager on a daily basis, to using a data terminal to report the hazard or create a work order. Regardless of the method used, the system must be available to all employees. Some workplaces that require the use of a data terminal have been found to have too few terminals for the employees to use or not all employees have been given access to the terminal system.

OSHA expects that employees will be able to report unsafe conditions anonymously if they so choose. This may be difficult if using only an electronic system. Electronic systems should be supplemented with a less formal suggestion box for employees who may not have access or wish to remain unidentified.

Employees must be encouraged to report unsafe conditions and not be discouraged to do so. The simplest form of encouragement is for management to say thank you for bringing the hazard to their attention. Other more elaborate rewards may include gifts of nominal value such as gasoline cards or cafeteria lunches. Efforts should be made to publically recognize employees for reporting unsafe conditions. There should also be a system for informing the employees of the report, along with a follow-up when the unsafe condition has been mitigated.

Employees may be discouraged when reported unsafe conditions are not corrected and the reporting employee is not informed of the reason. This results in an attitude of: Why should I report unsafe conditions when management does nothing about it? Another negative consequence of discouraging employees from reporting unsafe conditions is the possibility that they will contact OSHA to file a formal complaint that will likely result in an OSHA enforcement inspection.

Of course, employees will not be able to report unsafe conditions unless they have been trained to be able to recognize them. Like inspections, all employees should receive at least some level of hazard recognition training. Such training can be considered an investment. For a workplace with 200 production employees, training them all will result in an additional 200 workplace inspectors on the floor at all times to supplement the limited safety and health resources. Training the employees in hazard recognition techniques will also result in more detailed reporting of unsafe corrections and will reduce the number of frivolous reports.

Processing reported unsafe conditions should be similar to the processing of inspection findings. They should be promptly investigated and evaluated. Those that are found to not be valid unsafe conditions should be referred back to the reporting employee with an explanation of the reason the concern is not a hazard. Failure to report back to the employee will tend to discourage future reporting since the employee may feel that his or her concerns are simply being ignored. For those unsafe conditions that are determined to be valid, the conditions should be entered into a log and assigned appropriate corrective actions, with assignments for completion and completion dates. The corrective actions should also be reported back to the reporting employee to evidence managements' commitment to safety and health.

Employees must be protected from identified unsafe conditions without undue delay, even if the final correction may take some time. The immediate protection is usually referred to as an interim protection. Interim protection may take the form of:

- Personal protective equipment.
- Removing equipment from service.
- Barricades.
- Revised work practice.
- Administrative controls such as job rotation or more frequent breaks.

INVESTIGATION OF ALL INCIDENTS

To say the least, it is unfortunate when a safety- and health-related incident occurs that causes an injury, illness, or other damage to the facility or its employees. However, to allow a similar event to reoccur sometime in the future is truly tragic! It is extremely important that we learn from these types of events to identify not only what happened but also why it happened. Ideally, these determinations will result in corrective actions that will prevent similar occurrences in the future. It must be understood that if nothing is changed, that it is extremely likely that a similar event will occur again sometime in the future. Therefore, by definition, the only way of preventing a similar occurrence in the future is to see that appropriate changes in the form of corrective actions are implemented. The process to uncover the causes of incidents is a thorough and effective incident investigation.

It is critically important that an investigation be performed for all safety and health incidents that occur, even the most seemingly minor incidents like near misses. A near miss or as some say "near hit" is an event that could have caused an injury or property damage but simply through luck did not. While it may be true no injury occurred, that is no guarantee that a serious injury may not occur the next time. The formality of the investigation may differ based upon the seriousness of the incident being investigated and the potential for serious injury on a recurrence.

The incident investigation procedure should be written and include:

- Employees are required to report all incidents.
- Indicate when investigations must be conducted.
- Decide who will investigate the incidents.
- Under what circumstances are investigations conducted.
- Provide training for investigators.
- Implement root caused base analysis.
- Avoid routinely blaming employees.
- Track recommendations to correction.
- Review reports to ensure non recurrence.

Similar to reporting unsafe conditions, employees must be encouraged to report all incidents, regardless of how minor they may seem. Experience has shown that many employees will not report minor incidents for several reasons. One reason is they just do not want to be bothered with the process for what they consider just a minor incident such as a small cut or scratch. Another reason is they may have the perception that all incidents are considered the fault of the employee and result in a disciplinary action. Positive reinforcement must be used to counter these attitudes. Encouragement to report incidents may take the form of a positive recognition to a strong disciplinary action for failure to report the incident. Another way to overcome the reluctance of employees to report incidents is to properly investigate them and ferret out the true root cause and avoid the automatic placing of blame on the employee.

While all incidents should be investigated, the depth of the investigation may vary depending on the nature of the incident and the resulting consequences. An incident that results in a small paper cut does not have to be investigated with the same level of detail as one that results in an amputation on a piece of operational equipment. The latter incident should obviously receive a much greater degree of detail than the former. The paper cut investigation can simply be a report by the injured employee. The amputation would receive a much more detailed investigation including inspections of the equipment and reviews of the standard operating procedures (SOPs) as well as the associated job hazard analysis (JHA). The resultant investigation will likely include recommendations for improving the machine guards and revisions to the SOP and JHA.

The composition of the investigation team would also vary depending on the severity of the incident. As the severity increases, more people may join the investigation team, as either investigators or reviewers of the investigation reports. At a minimum for minor incidents, an investigation team should consist of the employee's supervisor and the injured employee. Additional information can be obtained from the witnesses. More sever incident investigation teams would include the safety and health manager, department manager, a representative of the safety committee, the maintenance manager, corporate safety and health staff, consultants, manufacturer representatives, and senior management.

All persons involved in incident investigations should receive training in incident investigation techniques. The training does not have to be at the same level for all those on the investigation team. The investigation team leader, who is usually the safety and health manager or the department supervisor, should have extensive incident investigation training. Others on the team, such as employees or the equipment or process experts, may be used as support and do not necessarily require the same level of training. The formal detailed training should include information on the principles of root cause analysis and the process used at the workplace. As with all other training, this training should be formally documented.

There are several techniques for incident investigations and several methodologies available. The typical investigation is based on determining the root cause of the incident. Too many incident investigations stop short of determining the true, or root, cause of the incident. The most common failure is to limit the investigation to the actions of the employee and the training they received. Root cause analysis is the methodology favored by OSHA and the VPP as it is most likely to discover all underlying causes of incidents.

It is widely believed that most incidents are the result of employee actions that are either contrary to the given procedure or the lack of attention to the process. Regardless of the truth or falseness of that belief, it does not mean that all incidents are the result of an employee's failure to follow rules or not paying attention. The fastest way to discourage employees from reporting an incident is to automatically place the blame of the incident on the employee. One very clear instance of this occurred at a VPP applicant that was undergoing the OSHA onsite evaluation. During the review of incident investigations, the reviewer observed that on the reverse side of most of them was the notation: "Until proven otherwise, it is determined that the employee was at fault." Clearly, the incident investigator, who was the department supervisor, was working under a mindset that the employee was at fault and further investigation would prove pointless. The employees at this particular work site were understandably concerned about reporting any incidents, especially considering that causing an incident would almost always result in some form of disciplinary action. The result of this finding by the OSHA evaluation team resulted in a recommendation of Merit with a goal of revising the incident investigation procedure to be more root cause based and that all investigators be trained to perform a root-cause-based investigation.

The results of all incident investigations must be fully documented in a written report. The report should identify the event and the conditions surrounding the event. The report should also include recommended corrective actions to avoid recurrences, and the corrective actions should be treated the same as those for hazards found during inspections or reported by employees. They must be assigned to a responsible person with a due date and tracked to their completion.

Unless the incident investigation reports are reviewed, there is no reason to complete the investigation. The investigation reports provide a very useful tool for determining the causes of incidents so that corrective measures can be taken to avoid recurrences. The reports should be reviewed for several reasons:

- Confirm that the investigations have been objective and detailed.
- Confirm that the investigations delved deep enough to uncover the true root causes of the incidents.
- Confirm that investigations did not unfairly place blame for incidents on employee misconduct or other factors.
- Review corrective actions to avoid recurrences.
- Revisions to existing procedures or programs have been made or new procedures or programs have been developed.
- Training on the new and revised procedures has been delivered.
- All employees have been made aware of the causes of the incidents.
- Recognition of trends in similar incidents are observed.

TREND ANALYSIS OF ALL SAFETY AND HEALTH INDICATORS

Throughout our lives and history, we can learn from our mistakes. Failure to learn from our mistakes and correct them or how we act in the future will result in our constantly repeating our history. One of the best tools we have to learn how employees may be hurt in our workplaces is to study how they have been hurt in the past. This type of study is referred to as a trend analysis of lagging indicators. The reason for the term *lagging indicators* is that since the event has already occurred, the review is lagging behind that action. Although that is not the best indicator, it is nonetheless a valuable one. The better indicators are those that have not resulted in incidents but may if they are not positive. These are referred to as *leading indicators*.

The purpose of performing a trend analysis is to determine the effectiveness of the safety and health management system using the history of prior events and activities. Using a trend analysis it can be determined if performance is acceptable requiring no change in operations or if the performance is declining and additional interventions are required.

Traditional trend analyses have focused their attention primarily on reviewing the following lagging indicators: the OSHA 200 and 300 logs, first reports of injuries, first-aid logs, near-miss reports, workers' compensation experience modification rates (EMR), and employee surveys. These indicators are useful in that they provide accurate information about the incident history. However, using only these indicators is not very effective to identify how employees may be injured or become ill in the workplace in the future. Since the results are historical, it is too late to prevent those injuries that have already occurred. Also, especially for a small company or one with very few data points to review, it would be difficult to determine any trends at all. Even when adding data from similar companies, there may not be enough data points to develop and useful information about what is causing most injuries. The primary problem with the historical nature of lagging indicators can be illustrated with an analogy to the stock market. The typical stock market disclaimer is something like "past history is no predictor of future earnings." In safety and health, the same holds true.

To address the problem with lagging indicators and to expand the usefulness of the trend analysis process, many companies have begun reviewing what is referred to as leading indicators. Those are the measures of selected criteria that may be used as predictors of future performance. If the selected performance indicators show poor or unacceptable performance, modifications or changes can be made before injuries actually occur. The focus on using leading indicators is more on the process of the safety and health management system rather than the end result.

There are numerous indicators that may be selected in developing a trend analysis based on leading indicators. There is no ideal formula for choosing those leading indicators that may be the best predictors of the effectiveness of the process. Instead each workplace would have to develop its own list of measures to include in its trend analysis process. The list would depend on a number of factors such as size of the workplace, number of employees, complexity of the workplace, previous injury history, and the sophistication of the data system.

A list of suggested data points, both leading and lagging, that may be considered in the development of a trend analysis process is provided in Exhibit 5.1. Although it is comprehensive, it should not be considered complete. We are sure that there may be other data points that may be very useful for some workplaces but not others.

Using some or all of the leading indicators in Exhibit 5.1, management should be able to identify those that are not meeting expectations and whose failure may eventually result in a deterioration of the effectiveness of the safety and health management system. That deterioration may then result in an unwanted, but not unexpected, negative incident such as an injury or property damage. Examples of the usefulness of some of the leading indicators include:

- *Percent near misses investigated* Not investigating all near-miss incidents is ignoring a valuable tool to identify the cause of the next serious incident.
- *Percent of new employees that have completed new employee orientation* Those employees who have not completed their safety and health orientation are more subject to injury and illness than those that have completed the training.
- *Percent of employees using PPE properly* A large percentage of employees not properly wearing PPE is an indication that they are more exposed to the hazards from which the PPE is supposed to protect them. It may also be an indication of a need to improve the training for the PPE use.
- *Percent of tasks with job safety analysis (JSA) conducted* If tasks are not reviewed for hazards, employees may be exposed to those unidentified hazards.
- *Percent preventative maintenance completed versus planned* Failure to maintain a current schedule may result in equipment breakdowns requiring expedited and unplanned repair or service. Such rushed work usually results in incidents due to inadequate planning and possible short-cuts taken to complete the work in the shortest possible time.

It must be emphasized that with even the most elaborate trend analysis process when making predictions, nothing is a sure thing. Lagging indicators of safety are accurate measures, but they provide the information too late to protect those that have already been affected. Leading indicators of safety are not absolute, but they can help identify weakness in the safety and health management system and be effective predictors of system failures to prevent injuries before they occur. The more leading indicators that are used the more effective the system will be in predicting the outcomes.

Exhibit 5.1 Leading and Lagging Safety and Health Indicators

Leading Indicators

Incidents

- Percent of reported injuries investigated
- Percent of reported injuries investigated by supervisors
- Percent identified corrective actions completed
- Percent identified corrective actions completed for reported concerns/complaints
- Percent near misses investigated
- Percent of safety related work or ders completed in 2 business days
- Percent of work place injuries reported within 2 days

Training

- Percent of training completed for staff that require safety training
- Percent of monthly safety talks completed
- Percent of new employees that have completed new employee orientation
- The effectiveness of the orientation and other training as determined by tests

Industrial Hygiene

- Percent of employees in hearing conservation program receiving annual audiogram
- Percent of employees in respiratory protection program that have a physician's written recommendation on file
- Percent of positions using hearing protection that have a noise exposure data on file

- Percent of positions using respirators that have an exposure data on file
- Percent of employees using PPE properly

Hazard Assessment

- Percent of planned safety audits completed
- Percent of identified safety hazards corrected within set period of time
- Percent of ergonomic evaluations conducted
- Percent of new employees' work stations evaluated
- Percent of tasks with job safety analysis (JSA) conducted
- Percent of equipment inspected that has guards in place and functional
- Percent of new operational changes that included hazard analysis

Policy/program

- Percent of safety behavior observations consistent with expectations
- Percent of employees involved in the safety and health management system
- Percent of time spent by supervisors and managers in safety and health activities
- Percent of scheduled safety committee meetings held
- Percent committee member attendance at safety committee meetings
- Percent of safety committee minutes published within 1 week
- Percent preventative maintenance completed versus planned

- Percent MSDS on hand for chemicals used
- Percent of MSDS received for new chemicals being used
- Number evacuation drills completed compared to required
- Percent safety policies reviewed annually
- Percent of fire extinguisher inspections in compliance
- Percent of organization's annual safety goals achieved
- Percent of position descriptions that outline health and safety responsibilities
- Percent of employee drivers with valid drivers' license
- Percent of positions that have physical demands identified

Lagging Indicators

- Percent of incidents reported
- Percent of injuries reported
- Percent of needle sticks reported
- Equipment damage reports
- OSHA recordable incident rates (TCIR and DART)
- OSHA days away from work
- Severity rate of recordable incidents
- Workers' Compensation claim rate (EMR)
- Workers' Compensation costs

6

HAZARD PREVENTION AND CONTROL

HIERARCHY OF CONTROLS

Once the occupational hazards have been identified in the work-site analysis phase, the next step is to address each of the hazards by controlling the safety and health hazards posed to the employees. VPP embraces the use of the hierarchy of controls to eliminate (or at least reduce) workplace hazards:

- Engineering controls
  - Physical changes to the workplace by the addition of equipment, such as ventilation and guards, or change in the process, such as substitution of less hazardous chemicals, can eliminate or reduce the safety and health hazards to employees.
- Administrative controls
  - Procedural steps can be taken during the task to reduce or eliminate safety and health hazards.
  - Providing additional training can eliminate or reduce employee exposure.
  - Use employee rotation to ensure chemical or physical agent permissible exposure limits are not exceeded.
- Personal protective equipment
- Equipment, such as respirators or gloves, are used to provide protection and eliminate or reduce employee exposure.

Preparing for OSHA's Voluntary Protection Programs. By Brian T. Bennett and Norman R. Deitch
Copyright © 2010 John Wiley & Sons, Inc.

Engineering controls are the most desirable choice as they usually eliminate the hazard altogether with little or no action from the employee. However, engineering controls are often expensive in both cost and time to design, install, and implement. Technical, maintenance, and operational personnel are often needed to fully analyze the system and then design an appropriate engineering control that will be effective in controlling the safety and health hazard, yet not hinder the ability of the equipment to operate as intended or the employee to perform the appropriate tasks.

The next most desirable control is the administrative control. Administrative controls may be successful in controlling or eliminating the hazard, so long as they are followed. Although the administrative control may be technically correct as written, it must be followed by the employee exactly as written in order to be effective. One reason administrative controls are not preferred is because they require voluntary employee input in order to be effective. An employee can receive the training or read the standard operating procedure, but the control is not effective if the employee chooses not to follow it explicitly.

Personal protective equipment is the least desirable of the controls since the employees need to make a conscious decision to use it in order to be protected from the hazard. Using this control does not eliminate or reduce the hazard; it just provides a barrier between the hazard and the employee.

Employees may choose not to wear personal protective equipment for various reasons, including:

- Employee does not perceive a need to wear it.
- Employee feels that the personal protective equipment hinders his or her ability to perform a task.
- Employee believes it takes longer to don the personal protective equipment than it does to perform the task.
- Environmental conditions are exacerbated by use of the personal protective equipment.
- The personal protective equipment does not fit properly and is uncomfortable to wear.

PROFESSIONAL EXPERTISE

As mentioned previously, the VPP does not mandate that companies employ certified safety and health professionals on their staff. It is acceptable to use certified safety and health professionals from the corporate staff, insurance companies, or consultants as required. In any case, the site must develop a plan as to how and when certified safety and health professional will be used, and how they will be used to enhance the effectiveness of the safety and health process.

It is highly recommended that VPP work sites have access to and use certified safety and health professionals when applicable. Certified safety and health professionals have met certain educational requirements (usually a baccalaureate degree), have many years experience (usually 5–7 years minimum) performing safety and health tasks full time, and have passed at least one comprehensive written examination to demonstrate a minimum base of knowledge and demonstrate competency in a particular discipline. Although there are knowledgeable and experienced safety and health professionals who are not certified, those that have gone through the certification process have demonstrated a minimum level of knowledge, experience, and competence in a particular field and hold a level of credibility as a recognized professional in their field of expertise.

The basic responsibility for the certified safety and health professional is to identify, prevent, and control recognized hazards in the workplace. Among the types of certified safety and health professionals that may be used include:

- *Occupational Medicine Professionals* This group may include physicians, nurses, audiologists, paramedics, and emergency medical technicians. These professionals may be responsible for the following types of activities:
  - Providing first aid to injured/ill employees.
  - Providing the appropriate therapy to ensure an employee can safely return to his or her normal job.
  - Administration of preventive programs, such as medical surveillance examinations and hearing and vision tests.
  - Administration of health wellness activities to help keep employees in good health, such as smoking cessation and weight control programs.
  - Participation in the review of work-site analysis and the development of standard operating procedures.
- *Industrial Hygienists* This group of professionals helps to ensure the health of employees by identifying and controlling occupational chemical and physical stressors. Included in this group are industrial hygienists, health physicists, and ergonomists. These professionals may be responsible for the following types of activities:
  - Identification, sampling for, and control of occupational chemical and physical health hazards. These hazards may include:
    - Chemicals
    - Dusts and fibers
    - Mists
    - Fumes
    - Noise
    - Heat or cold temperature extremes
    - Vibration
    - Ultraviolet or infrared light
    - Participation in the review of work-site analysis and the development of standard operating procedures.

- *Health Physicists* This group of professionals helps ensure the health of employees by identifying and controlling ionizing and nonionizing radiation hazards such as ultraviolet heat sealers, radio-frequency transmitters, X-ray machine, or nuclear measuring devices.
- *Ergonomists* This group of professionals helps ensure the health of employees by identifying and controlling hazards associated with the body's fit to their workplace. These professionals may be responsible for the following types of activities:
  - Evaluation of repetitive motion issues
  - Lifting loads
  - Bending, twisting, reaching
  - Jobs requiring long periods of sitting or standing
  - Evaluation of tools and their interface with the human body
  - Participation in the review of work-site analysis and the development of standard operating procedures
- *Safety Professionals* This group of professionals includes certified safety professionals (CSP), occupational health and safety technologists (OHST), construction health and safety technologists (CHST), and certified hazardous materials managers (CHMM). These professionals have received specialized training in the recognition and control of specific safety hazards in the workplace, such as confined spaces, lockout/tagout, scaffolding, fall protection, and chemicals. Safety professionals are typically involved in:
  - Work-site inspections
  - Work-site compliance audits
  - Employee training
  - Participation in the review of work-site analysis and the development of standard operating procedures
  - Developing, implementing, and managing the safety and health management system

A written plan must exist that will explain specifically how and when the site will request and utilize the services of these professionals. The plan should also include:

- What type of certified safety and health professionals are used in the evaluation and control of workplace safety and health issues
- Who the certified safety and health professionals are (internal site employee, corporate employee, or consultants)
- Who can request the use of the certified safety and health professionals (who is the site liaison to these resources?)
- How to contact the certified safety and health professionals (phone number, e-mail address)
- The field of expertise of the safety and health professionals

- The types of activities the certified safety and health professionals are involved with or may be involved with
- Who is responsible for coordinating and overseeing the activities of the certified safety and health professionals, including the receipt of recommendations, assignment of action items, and tracking action items to completion

It is possible that a VPP work site may use several safety and health professionals with various expertises at the same time to assist with the site's safety and health process. The work site must ensure that these professionals communicate with each other concerning the various tasks they are performing to ensure coordination and integration of their activities. Some VPP sites have found it beneficial to hold weekly meetings or conference calls with all involved parties to discuss the various activities being conducted to ensure everyone knows what is going on and activities are properly scheduled and completed. Finally, there must be a mechanism to ensure that any important safety and health communications are transmitted between the safety and health professionals and employees.

You should include a listing of the safety and health professionals you are using, along with their credentials and the various activities they are involved with in your VPP application.

SAFETY AND HEALTH RULES

The expectation is that a VPP work site will have a formalized, written safety and health process available for review by the VPP onsite evaluation team. As part of this process, various safety policies and procedures will be developed. There should be clear and concise policies and procedures to provide adequate employee protection for all of the hazards potentially present in the workplace, such as:

- Lockout/tagout
- Confined space entry
- Asbestos
- Hexavalent chromium
- Silica
- Powered industrial trucks
- Cranes
- Trenches and excavations

The safety and health procedures should be developed and reviewed periodically by a cross-functional team involving employees at all levels. The policies and procedures should be reviewed by a team annually to ensure they are up to date and reflect the current conditions in the work site. The procedures must also be reviewed by all affected employees so they know their roles and responsibilities in terms of implementing the policies and procedures. Ensure that this training is documented as it is something the VPP onsite evaluation team will review. The policies and procedures must also be available to all employees, at all times, either in writing or electronically.

An index of your policies and procedures must be attached to your VPP application.

SAFETY AND HEALTH DISCIPLINE

A VPP work site must also have a disciplinary system in place for violations of safety and health policies and procedures. This program is separate from the disciplinary system used for contractors. The disciplinary program must outline how employees will be dealt with in the event of noncompliance with the established safety and health policies and procedures. The disciplinary system must apply equally to salaried and hourly workers. During the course of your VPP onsite evaluation, the team will review the discipline that has been issued for safety and health violations. The team must be comfortable that the policies and procedures are equally enforced to all employees, regardless of their organizational rank. In work sites where safety and health has become well ingrained and established as a core value, discipline for safety and health violations should be far and few between. However, the expectation is that some employees will have been disciplined at some point over the past few years. Another measure of the effectiveness of the discipline program is the level of awareness that the employees have of it.

The typical disciplinary system for safety and health violations is progressive in nature. The first infraction results in a verbal warning, the second infraction results in a written warning, the third infraction results in a suspension, and the fourth violation results in a termination of employment. It is a good practice to document all disciplinary actions, even a verbal warning. Verbal warnings should be documented for two main reasons: first, to memorialize the discussion, and second, to provide a record as most discipline at VPP sites does not go beyond the verbal warning step.

The progression of discipline may skip a step or a number of steps depending on the severity of the safety and health infraction. There must also be a mechanism to inform employees about the safety and health disciplinary system, including the penalties that may be levied.

Safety and health performance should not result in only negative reinforcement. There should also be a positive reinforcement system for those employees who demonstrate outstanding safety and health performance. You should consider a system to recognize situations when employees have gone above and beyond the minimum safety and health requirements. It is common to find VPP sites recognizing employees for the following types of activities:

- Participation in hazard analysis teams
- Participation in safety and health policy and procedure review teams
- Presenting safety and health training
- Conducting safety inspections

Care must be taken in establishing the recognition system. The award itself should not have a significant intrinsic value, but rather a nominal award designed more to recognize and reward good performance. Typically, these nominal awards include things such as free lunches, clothing items, or gift certificates for local merchants.

Work sites should avoid rewarding employees for zero injuries or accidents. Experience has found that this type of incentive, especially if it has significant value, may influence employees to not report injuries or incidents in order to receive the award. Although it is everyone's goal to avoid injuries and incidents, a more appropriate goal may be to recognize milestones such as one year without an away-from-work injury or the completion of significant goals such as achieving OSHA VPP status.

You are required to provide a description of any positive reinforcement programs you have in place to recognize safety and health performance in your VPP application.

PERSONAL PROTECTIVE EQUIPMENT

Although personal protective equipment is the least desirable method of protecting employees from recognized hazards, it will still be used as part of your hazard control program. Care must be taken in selecting personal protective equipment. A cross-functional team involving all levels of employees should be assembled to develop the personal protective equipment (PPE) program. The first step is to conduct a PPE assessment. This is done by conducting a thorough evaluation of the various jobs and tasks and identifying all of the chemical and physical hazards to which an employee may be potentially exposed. A checklist should be developed for the PPE assessment to ensure they are all complete and consistent.

Once the hazards have been identified, a plan can be developed to identify the appropriate PPE to protect the employees. Once again, employees should participate in the PPE selection process. The first step in selecting appropriate PPE is to review chemical compatibility charts and specification sheets to ensure the PPE will provide the necessary protection. Once the appropriate type of PPE has been selected, PPE samples can be obtained from vendors (usually at no cost) for employees to test. This step is important because you want to ensure that the PPE fits properly, is comfortable to wear, and provides the level of protection necessary before a significant investment is made. If employees are actively involved in this trial step, it is much more likely they will wear it. It also provides one more opportunity for them to be actively involved in the safety and health process.

This PPE assessment should be revalidated periodically so that any changes to the work site, equipment, or job task can be evaluated.

Once the PPE assessment is complete and the appropriate PPE purchased, employee training must be conducted before the PPE is issued to employees for use. Training topics include:

- *Selection of PPE* Employees must be trained on the proper selection of PPE, including which PPE should be used to protect against each of the hazards that may be present in the workplace. Employees must know the specific PPE requirements, for example, the correct glove material [latex, poly vinyl chloride (PVC), etc.] and the correct respirator cartridge (organic vapor, HEPA, etc.).

- *Distribution of PPE* Employees must be aware of how to obtain the necessary personal protective equipment that may be needed. Obtaining PPE on off shifts on weekends must also be addressed. The general rule of thumb is PPE must be readily available at all times for immediate use by the employee. Employees who distribute the PPE should be trained on how to properly size PPE for employees and any other specific requirements such as fit testing for respirators.
- *Donning of PPE* Employees must be familiar with how to properly put on and take off PPE, including appropriate disposal methods for used PPE. Employees must be knowledgeable about how to ensure they have selected the proper size of PPE. Employees also need to know how to test the PPE for proper donning (such as an air-purifying respirator positive and negative seal check) before being exposed to a hazard.
- *Inspection of PPE* Employees must know how to conduct a preuse and postuse inspection of personal protective equipment, including what to do if the PPE fails the inspection. The employees must be educated on the conditions that warrant taking the PPE out of service, and the proper method of doing so to ensure that defective PPE is not used by someone else.
- *Use of PPE* Employees must be trained on how to properly use the personal protective equipment so that the maximum level of protection is provided. They must also be familiar with when PPE must be used for the various job tasks they are responsible for conducting.
- *Maintenance of PPE* Maintenance and repair of personal protective equipment should be limited only to those employees or manufacturer's representatives who have had specialized training to make repairs or modifications. Employees who do not have this specialized training should not be allowed to make repairs or modifications, but rather know what to do with damaged PPE. Many work sites avoid the issue of PPE maintenance by using disposal PPE.

Some work sites may require the use of air-supplied or air-purifying respiratory protection. Your VPP application must provide a description of the various types of respirators that are used, and what contaminants they protect against. If respirator use is required, the site must have:

- A written respiratory protection program
- An employee medical surveillance program
- A respirator training program
- A respirator fit testing program

The written respiratory protection program must be complete, updated in the past 12 months to reflect the current conditions, and comply with 29 CFR 1910.134. The written program, or at least the table of contents, must be attached to your VPP application.

As a minimum, the medical questionnaire required by 29 CFR 1910.134 must be completed by all employees required to use a respirator. Quite often VPP sites go beyond the minimum requirements of the OSHA standard and require an annual medical exam including a pulmonary function test to better protect the employee's well-being.

Employees must be trained annually on all aspects of the respiratory protection program, including the proper selection, use, end of service life, and maintenance of respirators.

Finally, a fit test must be completed annually for all employees required to use respiratory protection. Although qualitative fit testing is allowable under the OSHA standard, most VPP sites perform quantitative fit testing as it is more likely to ensure a proper fit of the face piece and hence will ensure a higher level of protection for the employee.

EMERGENCY PREPAREDNESS

In the post-9/11 world, emergency preparedness and response plays a very important part in your overall safety and health process. Regardless of what type activities are performed at your work site, you must have an appropriate emergency response plan, it must be written, and employees must be trained on their roles and responsibilities.

Emergency planning starts with a thorough survey of the work site, identifying all of the potential emergencies that may arise. Included in this analysis should be all accidental, natural, and intentional scenarios such as:

- Accidental
  - Fire
  - Explosion
  - Medical emergency
  - Rescue (confined space, trench collapse, etc.)
  - Hazardous materials release
- Natural
  - Flood
  - Earthquake
  - High winds/tornado/hurricane
- Intentional
  - Sabotage
  - Terrorist attack

Once each of the potential scenarios has been identified, the site should develop written preemergency plans for each. A preemergency plan details the specific action that should be taken by site employees and emergency responders to successfully mitigate the emergency situation. The advantage of preparing a preemergency plan is that a well-rounded group of experts can analyze the situation in a nonstressful (e.g., nonemergency) environment and develop options to safely, efficiently, and effectively mitigate the situation. It also provides an opportunity to identify any specialized training that may be required for employees and responders, as well as any specialized equipment that may be needed. These preemergency plans should be periodically reviewed for completeness and practiced as part of the site's emergency response training and exercise program.

Although there is no requirement for a work site to have an emergency response team, VPP work sites often do have them. The emergency response team should be developed to address the various emergency scenarios that have been identified previously in the emergency planning process. The basic mission of the emergency response team should be to:

- Protect the lives of employees, contractors, and the community
- Protection of the environment
- Protection of company property

Typically, emergency response teams provide one or more of the following services:

- Fire fighting
- Hazardous materials response
- Emergency medical services
- Rescue

Each site must thoroughly review its potential emergency scenarios and ascertain what services are available from the local governmental emergency responders. If the local responders do not have the proper equipment or training to provide the necessary emergency services, or their response times are not adequate, then a work site emergency response team may be warranted. Ultimately, each work site will need to determine what level of response is necessary to adequately protect all employees. For example, an office building in a city may only require a first-aid team, whereas a chemical plant located in a rural area may require fire fighting, emergency medical, hazardous materials, and a rescue team.

Once the type of emergency response team has been determined, a mission statement that clearly delineates the response team's roles and responsibilities should be developed. An appropriate emergency response training program that matches the mission statement must be developed and implemented. The training should be a comprehensive blend of classroom and hands on training ultimately leading to employee certification. Employee certification should be to national consensus standards, such as the National Fire Protection Association (NFPA) for fire fighting, hazardous materials, and rescue and the National Registry of Emergency Medical Technicians (NREMT), the National Safety Council, American Heart Association, and American Red Cross for emergency medical services.

The training curriculum should also follow the national consensus standards and OSHA standards to ensure all required learning objectives have been met. The VPP onsite evaluation team will review the initial certification curriculum as well as the annual recertification training requirements.

Appropriate incident management system training for key personnel is also required by OSHA and NFPA standards as well as the National Incident Management System. Key personnel should be identified, and each incident management system filled with a primary and backup employee. Personnel with incident management system responsibilities should complete Department of Homeland Security Federal Emergency Management Agency's (FEMA) National Incident Management System (NIMS) training curriculum.

VPP sites typically conduct an annual emergency exercise involving site personnel as well as the local emergency responders. As a minimum, the VPP onsite evaluation team will ensure that each and every employee has participated in an emergency evacuation drill in the past 12 months. These drills should be conducted at various times of the workday to ensure that all employees on all shifts have the opportunity to participate in the drill. It is best if the evacuation drills are conducted on all shifts and weekends. There must be provisions to allow employees who miss the initial training to participate in a makeup session.

Appropriate emergency response equipment must also be onsite. The equipment should be commensurate to the training level of the site's emergency responders. For example, if the first-aid team is trained on the use of an automated external defibrillator (AED) and the delivery of oxygen, then the expectation would be that the work site would have an AED and oxygen resuscitator.

All emergency response equipment must be maintained in good working condition and inspected per OSHA or consensus standards and the manufacturers recommendations. A training program for those responsible for inspecting emergency equipment should be developed, including written procedures on how to conduct the inspection and the maintenance of written records documenting the inspection results. The inspection records should be filed and readily retrievable for review by the VPP onsite evaluation team.

A critical part of any emergency preparedness and response program is emergency exercises. The scenarios that are chosen for emergency drills and exercises should be selected from the various emergency scenarios and emergency preplans that have been developed previously. The purpose of the emergency exercise program is to test the effectiveness of the preemergency response plans, the training program, communications systems, and coordination with outside agencies. VPP work sites are expected to complete at least one full-scale emergency exercise involving the municipal emergency responders annually.

At the conclusion of the emergency drill or exercise, a critique should be conducted to identify all of the things that worked well and those areas where some improvements can be made. Once the opportunities for improvement have been identified, they should be investigated to see if changes are warranted. If so, the appropriate programs should be modified and retested.

PREVENTATIVE MAINTENANCE

Equipment that is not properly maintained can lead to injuries, leaks of dangerous materials, lost product, and poor quality. The VPP onsite evaluation team will check various pieces of equipment, especially safety systems such as safety showers and eyewashes, to ensure equipment is in good shape, the proper inspections are being conducted, appropriate records are maintained, and employees have been trained in the proper maintenance techniques.

There are two systems that can be used to ensure equipment is maintained in proper operating condition:

- Preventative maintenance
- Predictive maintenance

Preventative maintenance is the practice of ensuring the equipment is in a safe condition to operate productively and economically until the next inspection. Preventative maintenance also calls for the necessary mechanical action to be taken to restore the equipment to a safe condition. There are a number of forms of preventative maintenance, such as all pumps within the facility will be brought in to the shop every 6 months for servicing.

Predictive maintenance infers predicting potential problems, and the advisable or necessary preventative maintenance action necessary to avoid damage or failure. The disadvantage with this system is that while some equipment may have worked 2000 hours, others may have worked only 50 hours. So it is unnecessary, and therefore expensive, to remove, disassemble, reassemble, and replace equipment in terms of labor costs.

The difference between preventive and predictive maintenance is:

- Preventive solutions establish a calendar-based routine service event based on historical data.
- Predictive solutions can either support a preventive maintenance program or operate as a standalone solution that monitors critical-based equipment in real time.

While the VPP does not require a facility use either preventative or predictive maintenance, it does require a suitable system be in place to ensure employee safety. It is up to each individual facility to decide whether to use one or both of these programs to meet the VPP requirements.

The VPP onsite evaluation team will be particularly interested in how the frequency of inspections were determined. The appropriate frequency of inspections is not randomly selected; it should be determined by consulting various technical documents, such as:

- OSHA standards
- American Petroleum Institute (API) standards

- National Fire Protection Association (NFPA) standards
- American Society for Testing and Materials (ASTM) standards
- International Standards Organization (ISO)
- Manufacturer's guidelines

The expectation is there will also be an automated system in place to ensure that all inspections are conducted on time at the appropriate frequency. Included in this system should be a procedure outlining how to handle equipment that is not inspected on time.

PROCESS SAFETY MANAGEMENT

OSHA's Process Safety Management (PSM) standard, 29 CFR 1910.119, has received renewed interest in the past several years due to some highly visible catastrophes that have occurred. Sites that use a listed chemical above a threshold quantity are covered by the PSM standard. The VPP has created an application supplement for those facilities that are covered by the standard. See application in Exhibit 6.1.

In addition, there is an expanded annual evaluation requirement for PSM-regulated sites that will be covered in Chapter 9.

MEDICAL PROGRAMS

Health care providers and emergency responders should be afforded the opportunity to tour the facility at least once per year so that they are familiar with the operations and the potential hazards. Some basic instruction on the equipment and hazardous materials used onsite will be helpful in these professionals being able to properly recognize and treat injuries and illness and early on limit the severity of harm.

The site's medical programs must be fully integrated with the overall safety and health management system. The medical program can be broken down into two categories: the availability of onsite medical services and the availability of offsite medical services.

Onsite medical services may be provided by a physician, nurse, practitioner, or emergency responder. These services should be provided by an appropriately licensed or certified professional. Typically, physicians, nurses, and practitioners provide the following type of services:

- Medical surveillance
  - Either baseline preemployment physical examinations or routine, periodic examinations specified in either OSHA standards or internal procedures
- Follow-up treatment of injured or ill employees to ensure a safe return to work
- Workers' compensation case management
- Treatment of injured/ill employees

The physician, nurse, or practitioner may be a full-time or part-time employee or a contractor. They can have facilities either on or offsite. If a contractor is used, there must be provisions for 24-hour service, if the facility operates 24 hours per day.

Typically, emergency responders provide the following services:

- *Rescue* Extrication and removal of an injured employee from an area of danger to a safe area
- *Treatment* Providing first aid, CPR, or external defibrillation of an injured employee

As mentioned previously, emergency responders must be appropriately certified by a recognized agency and properly equipped to do their job.

Emergency medical services must be available for all shifts. If the site will depend on municipal emergency responders instead of an internal team, there should be an agreement that emergency services, as well as emergency transport to a hospital, will be provided on a 24-hour schedule. Coordination with a local clinic or hospital to accept injured employees should also be performed.

It is helpful to involve professional health care providers in the routine hazard analysis of the work site. They can bring a unique perspective into the various potential health hazards that may be present, and they will be invaluable in helping to develop appropriate hazard controls.

SAFETY AND HEALTH PROGRAMS

The site should list any significant engineering and administrative controls in its VPP application. A list of all significant written policies and procedures (and work permits, if any) should be listed, including the following (where applicable):

- Asbestos program
- Bloodborne pathogens
- Burning, welding, and hot-work program
- Chemical hygiene plan
- Confined-space entry program
- Crane program
- Fall protection program
- Hazard communication program
- Hearing conservation program
- Lead program
- Line break/line entry program
- Lockout/tagout program
- Management of change program

- Powered industrial truck program
- Radiation program
- Scaffolding program
- Hand and power tool program
- Trench and excavation program

Exhibit 6.1 VPP Application Supplement for Sites Subject to the Process Safety Management (PSM) Standard

The VPP Application Supplement for Sites Subject to the Process Safety Management (PSM) Standard has been submitted to the Office of Management and Budget (OMB) for review under the Paperwork Reduction Act of 1995. No person may be required to respond to, or may be subject to a penalty for failure to comply with, this supplement until it has been approved.

The public may submit comments on this application as well as other VPP information collection requirements at http://www.reginfo.gov. You may also obtain an electronic copy of the complete Voluntary Protection Information Collection Request (ICR) at this website. Click on "Inventory of Approved Information Collections, Collections Under Review, Recently Approved/ Expired," then scroll under "Currently Under Review" to Department of Labor (DOL) to view all of the DOL's ICRs, including those ICRs submitted for extensions. To make inquiries, or to request other information, contact Mr. Todd Owen, OSHA, Directorate of Standards and Guidance, Room N-3609, U.S. Department of Labor, 200 Constitution Avenue, NW., Washington, DC 20210; telephone (202) 693-2222.

VPP Application Supplement for

Sites Subject to the Process Safety Management (PSM) Standard

VPP applicants whose operations are covered by the Process Safety Management (PSM) Standard must provide responses to each question that is applicable to their operations. Responses must cover all PSM-related operations. Please indicate that a question is "Not Applicable" if it addresses functionality outside the scope of the operations, and briefly explain why.

I. Management of Change

A. Has the throughput changed from its original design rate? Has the site conducted a management of change (MOC) procedure for each throughput change since May 26, 1992?

B. For the MOC procedures conducted for the unit(s), has the procedure listed the technical basis for the change and ALL potential safety and health impacts of the change prior to its implementation?

C. From the site's list of MOCs, identify the oldest MOC procedure which might affect the integrity of one or more pressure vessels in the unit(s). Do these MOC procedures meet all 1910.119(l) requirements?

D. Does the MOC process address temporary changes as well as permanent changes?

E. Have MOCs been conducted on all changes to process chemicals, technology, equipment and procedures, and changes to facilities that affect a covered process?

II. Relief Design

A. For each throughput MOC procedure conducted, has the procedure addressed a review/analysis of the relief system (includes relief devices, relief discharge lines, relief disposal equipment and flare system) to determine if there may be any safety and health impacts due to increased flow as a result of throughput changes which might impact the existing relief system?

Guidance: An MOC procedure is required anytime a change per the requirements of 1910.119(l) is considered. An MOC procedure is a proactive management system tool used in part to determine if a change might result in safety and health impacts. OSHA's MOC requirement is prospective. The standard requires that an MOC procedure be completed, regardless of whether any safety and health impacts will actually be realized by the change.

B. After a change in the throughput in the unit(s), did the process hazard analysis (PHA) team consider the adequacy of the existing relief system design with respect to the increased throughput during the next PHA?

Guidance: Typically, the PHA team does not do a relief system engineering analysis. However, the PHA team should determine, through proper evaluation and consultation with the engineering/technical staff, if the existing/current engineering analysis of the relief system is adequate for the current/actual unit throughput.

If the throughput change was implemented between the time the PSM standard became effective (May 26, 1992) and the time the original PHA was required based on the PHA phase-in schedule, the original PHA would need to address the throughput change. However, if there was a throughput change after the original PHA, the next PHA update/"redo" or PHA revalidation would need to address the throughput change. In either event, an MOC procedure on the throughput change would need to have been conducted and incorporated into the next scheduled PHA.

C. Does the site's process safety information (PSI) include the codes and standards used in the design of relief systems?

D. Does the site's PSI include the relief system design and design basis?

Guidance: This includes the original design and design changes. Examples of PSI related to relief devices, their design and design basis include, but are not limited to such items as:

1. *Identification/descriptor of each relief device;*
2. *A listing of all equipment which will be relieved through the device;*
3. *Design pressure;*
4. *Set pressure;*
5. *Listing of all sources of overpressure considered;*
6. *Identification of the worst case overpressure scenario or relief design;*
7. *State of material being relieved (i.e., liquid, vapor, liquid-vapor, liquid-vapor-solid, along with an identification of the material which was the basis for the relief device selection);*
8. *Physical properties of the relieved materials, vapor rate, molecular weight, maximum relieving pressure, heat of vaporization, specific gravity and viscosity; and*
9. *Design calculations.*

Similar design and design bases PSI are required for the rest of the relief system equipment downstream from the relief devices, i.e., relief vent lines, manifolds, headers, other relief disposal equipment, and flare stack.

E. Are there intervening valves on the upstream or downstream lines to/from relief devices? If so, does the PHA consider the possibility that these valves could be closed during operation, rendering the relief devices non-functional?

F. If there are intervening valves on the upstream or downstream lines to/from relief devices, does the site have effective controls in place to ensure these intervening valves remain open during operations?

G. If there are intervening valves on the upstream or downstream lines to/from relief devices, is there an administrative procedure (e.g., car-seal procedure) to assure these valves are in the open position during operations? If so, has this procedure been subsequently audited?

H. Are there open vents which discharge to atmosphere from relief devices? If so, has the PHA considered whether these relief devices discharge to a safe location?

Guidance: PHA teams must address basic questions regarding what happens to the hazardous materials after they are relieved to atmosphere, including:

1. *Are there negative effects on employees or other equipment that could cause another release ("domino effects") of hazardous materials/HHC?*

2. *What presumptions or assessments exist to support that there will be no negative effects of an atmospheric release of hazardous materials/HHC?*

3. *Are employees near where relief devices discharge, including downwind locations (e.g., on the ground, on platforms on pressure vessels in the vicinity of elevated relief devices, etc.)?*

4. *Could a release from a relief device cause a release from other equipment, or could other nearby equipment affect the released material (e.g., a furnace stack could be an ignition source if it is located proximate to an elevated relief device that is designed to relieve flammable materials)?*

Part of the site's PHA team's evaluation, after it identifies the locations of open vents, is to determine if employees might be exposed when hazardous materials are relieved. If the PHA team concludes that a current and appropriate evaluation (such as the use of dispersion modeling) has been conducted, the evaluation could find that the vessels/vents relieve to a safe location. If the PHA team determines that this hazard has not been appropriately evaluated, the PHA team must request that such an evaluation be conducted, or make some other appropriate recommendation to ensure that the identified hazard/deviation is adequately addressed.

I. Does the site have a mechanical integrity (MI) procedure for inspecting, testing, maintaining, and repairing relief devices which maintains the ongoing integrity of process equipment?

J. Does the process use flares? If so, verify that the flares have been in-service/operational when the process has been running. If the flares have not been in-service, has the site used other effective measures to relieve equipment in the event of an upset? Has an MOC procedure been used to evaluate these changes?

III. <u>Vessels</u>

A. Do pressure vessels which have integrally bonded liners, such as strip lining or plate lining, have an MI procedure which requires that the next scheduled inspection after an on-stream inspection be an internal inspection?

B. Does the site have an MI procedure for establishing thickness measurement locations (TML) in pressure vessels, and does the site implement that procedure when establishing the TML?

C. Does the site have an MI procedure for inspecting pressure vessels for corrosion-under-insulation (CUI), and does the site inspect pressure vessels for CUI?

D. Does the site's MI procedure address testing (e.g., leak testing) and repair of pressure vessels? For example, does the MI procedure indicate how the testing and repair will be conducted and which personnel are authorized to do the testing and repair, including what credentials those conducting the testing and repair must have?

Guidance: API 510 requires in-service pressure vessel tests when the API authorized pressure vessel inspector believes they are necessary.

Guidance: Recognized and Generally Accepted Good Engineering Practices (RAGAGEP) that require credentials include, but are not limited to:

1. *Credentials for pressure vessel inspectors, see API 510, Section 4.2.*

2. *RAGAGEP for pressure vessel examiners credentials/experience and training requirements, see API 510, Section 3.18.*

3. *RAGAGEP for contractors performing NDE are the training and certification requirements ASNT-TC-1A, see CCPS, Section 10.3.2.1, (In-service Inspection and Testing) Nondestructive Examination.*

4. *RAGAGEP for qualifications for personnel who conduct pressure vessel repairs, alteration and rerating including qualifications for welders, see API 510, Section 7.2.1 and the BPVC, Section IX.*

5. *RAGAGEP for certifications at CCPS, Section 5.4 Certifications, Table 5.3, Widely Accepted MI Certifications, and Table 9.13, Mechanical Integrity Activities for Pressure Vessels.*

E. Were any deficiencies found during pressure vessel inspections? If so, how were they resolved?

Guidance: A deficiency (as per 1910.119 (j)(5)) means a condition in equipment or systems that is outside of acceptable PSI limits. In the case of a pressure vessel, this could mean degradation in the equipment/system exceeding the equipment's acceptable limits (e.g., operating a vessel, tank or piping with a wall thickness less than its retirement thickness).

F. Do the operating procedures for pressure vessels list the safety systems that are applicable to the vessels?

Guidance: Examples of safety systems include but are not limited to: emergency relief systems including relief devices, disposal systems and flares; automatic depressurization valves; remote isolation capabilities, aka emergency isolation valves; safety-instrumented-systems (SIS) including emergency shutdown systems and safety interlock systems; fire detection and protection systems; deluge systems; fixed combustible gas and fire detection system; safety critical alarms and instrumentation; uninterruptible power supply; dikes; etc.

G. Have there been any changes to pressure vessels or other equipment changes that could affect pressure vessel integrity, such as a change to more corrosive feed, a change in the type of flange seal material used or the vessel heads or nozzles, etc.,? If so, was an MOC procedure completed prior to implementing the change?

IV. Piping

A. Is there information in the MI piping inspection procedures or other PSI that indicates the original thickness measurements for all piping sections?

B. Is there information in the MI piping inspection procedures or other PSI that indicates the locations, dates and results of all subsequent thickness measurements?

C. Is there anomalous data that has not been resolved for any piping? (For example, the current thickness reading for a TML indicates the pipe wall thickness is greater/thicker than the previous reading(s) with no other explanation as to how this might occur.)

D. Has each product piping been classified according to the consequences of its failure?

Guidance: If the site inspects and tests all piping the same, regardless of the consequence of failure of the piping (i.e., piping inspections are implemented using the same MI program (1910.119(j)(2) and action/task (1910.119(j)(4) procedure for all piping without consideration of their consequence of failure or other operational criteria), then this question is not applicable.

E. Based on a review of piping inspection records, have all identified piping deficiencies been addressed?

Guidance: An example of a piping deficiency would be a situation where piping inspection data indicates that its actual wall thickness is less than its retirement thickness, and the site has conducted no other evaluation to determine if the piping is safe for continued operation. For a discussion on equipment deficiencies the definition of deficient/deficiency.

F. How does the site ensure that replacement piping is suitable for its process application?

Guidance: Typically, piping replacements are replacements-in-kind (RIK) when the process service does not change. However, if the piping replacement is not an RIK, then an MOC procedure is required.

G. Does the site's MI procedure list required piping inspectors' qualifications, welders' qualifications for welding on process piping, and when qualified welding procedures are required?

H. Is there information in the MI piping inspection procedures or other PSI that indicates the original installation date for each section of piping?

I. Is there information in the MI piping inspection procedures or other PSI that indicates the specifications, including the materials of construction and strength levels for each section of piping?

J. Does the site's MI procedure for piping inspections list criteria/steps to be followed when establishing TML for injection points in piping circuits?

V. Operating Procedures – Normal Operating Procedures (NOP), Emergency Shutdown Procedures (ESP) and Emergency Operations (EOP)

A. Are there established operating procedures, including: normal operating procedures (NOP), emergency operating procedures (EOP), and emergency shutdown procedures (ESP)?

B. Are operating procedures implemented as written?

C. Are there ESP for the all Unit(s), and if so, do these ESP specify the conditions that require an emergency shutdown?

Guidance: ESP are usually warranted during events that may include the failure of process equipment (e.g., vessels, piping, pumps, etc.) to contain or control HHC releases, loss of electrical power, loss of instrumentation or cooling, fire, explosion, etc. When EOP do not succeed during upset or emergency conditions in returning the process to a safe state, implementation of an ESP may be necessary.

When normal operating limits for parameters such as pressure, temperature, level, etc., are exceeded during an excursion, system upset, abnormal operation, etc., a catastrophic release can occur if appropriate actions are not taken. These actions must be listed in the EOP and must specify the initiating conditions or the operating limits for the EOP (e.g., temperature exceeds 225°F or pressure drops below 15 psig).

Information typically listed in EOP and/or ESP includes, but is not limited to the responsibilities for performing actions during an emergency, required PPE, additional hazards not present during normal operations, consequences of operating outside operating limits, steps to shutdown the involved process in the safest, most direct manner, conditions when operators must invoke the emergency response plan, or scenarios when they themselves must stop and evacuate.

D. Have control board operators received sufficient training, initial and refresher, to be qualified to shutdown the units?

E. Does the ESP specify that qualified operators are assigned authority to shutdown the unit(s)?

F. Are qualified control board operators authorized or permitted to initiate an emergency shutdown of the unit without prior approval?

G. Do EOP procedures identify the "entry point," i.e., the initiating/triggering conditions or operating limits when the EOP is required, the consequences of a deviation from the EOP, and the steps required to correct a deviation/upset once the operating limits of the EOP have been exceeded?

H. Do NOP list the normal operating limits or "exit points" from NOP to EOP; the steps operators should take to avoid deviations/upsets; and the precautions necessary to prevent exposures, including engineering and administrative controls and PPE?

Guidance: For NOP, the "operating limits" required are those operating parameters that if they exceed the normal range or operating limits, a system upset or abnormal operating condition would occur which could lead to operation outside the design limitsof the equipment/process and subsequent potential release. These operating parameters must be determined by the site and can include, but are not limited to, pressure, temperature, flow, level,composition, pH, vibration, rate of reaction, contaminants, utility failure, etc.

It is at the point of operation outside these NOP "operating limits" that EOP procedures must be initiated. There may be a troubleshooting area defined by the site's EOP where operator action can be used to bring the system upset back into normal operating limits. During this troubleshooting phase, if an operating parameter reaches a specified level and the process control strategy includes automatic controls, other safety devices (e.g., safety valves or rupture disks) or automatic protection systems (e.g., safety instrumented systems/emergency shutdown systems), would activate per the process design to bring the process back to a safe state. Typically, once the predefined limits for troubleshooting have been reached for a particular operating parameter, the process has reached a "never exceed limit". A buffer zone is typically provided above (and below if applicable) the trouble shooting zone ("never exceed limit") to ensure the operating parameters do not reach the design safe upper or lower limit of the equipment/process. This design safe upper and lower limits of the equipment or process are also known as the boundaries of the design operating envelope or the limit above (or below) which it is considered unknown or unsafe to operate. Once the operating parameter(s) reach the buffer zone entry point, there is no designed or intentional operator intervention (i.e., troubleshooting) to bring the process system upset back to a safe state. Any intervention in the buffer zone is as a result of the continued activation of the safety devices and automatic protection systems which initially activated at the predefined level during the troubleshooting phase. All of these predefined limits are important information for operators to know and understand and must be included in the PSI and operating procedures.

I. Are operating procedures implemented as written?

VI. <u>PHA, Incident Investigation, and Compliance Audits Findings/Recommendations</u>

A. Have all corrective actions from PHA, incident investigations, MOCs, and compliance audits been corrected in a timely manner and documented? Provide a list of all outstanding corrective actions, the date of corrective initiation, and the projected completion dates.

Guidance: There may be instances when a PHA team identifies deficiencies in equipment/systems which would violate the requirements of 119(j)(5) if left uncorrected. If the site continues to operate the deficient equipment/system, they must take interim measures per 119(j)(5) to assure safe operation, and they must also meet the 119(e)(5) requirements to resolve the findings and recommendations related to the identified deficiency.

The phrase from 119(j)(5), "safe and timely manner when necessary means are taken to assure safe operation", when taken in conjunction with 119(e)(5) means that when a PHA team identifies a deficiency in equipment/systems and the site does not correct the deficiency before further use, the site's system for promptly addressing the PHA team's findings and recommendations must assure: 1) that the recommendations are resolved in a timely manner and that the resolutions are documented; 2) the site has documented what actions are to be taken, not only to resolve the recommendation, but to assure safe operation until the deficiency can be corrected; 3) that the site complete actions as soon as possible; and 4) that the site has developed a written schedule describing when corrective actions related to the resolution and any interim measures to assure safe operations will be completed.

The system that promptly addresses and resolves findings and recommendations referred to in both 1910.119(e)(5) and 1910.119(m)(5) are not *requirements to develop a management program for globally addressing the resolution of findings and recommendations. Rather, these "system" requirements address how each specific finding and recommendation will be individually resolved (Hazard Tracking requirement under VPP). Each finding or recommendation will have its own unique resolution based on its nature and complexity.*

B. Has the PHA incorporated all the previous incidents since May 26, 1992 which had a likely potential for catastrophic consequences?

VII. Facility Siting/Human Factors

A. Does the PHA consider the siting of all occupied structures? *Guidance: Facility siting considerations for occupied structures include both permanent and temporary (e.g., trailers) structures.*

Global/generic facility siting questionnaires/checklists. Some employers (PHA teams) attempt to comply with this 1910.119(e)(3)(v) requirement by answering global/generic facility siting questions on a short questionnaire/checklist. PSM is a performance standard and the means the site uses to *comply with the standard are generally up to them as long as their performance ensures compliance with the requirement of the standard. If the site uses a questionnaire/checklist as part of its PHA to identify, evaluate and control all hazards associated with facility siting, this is permissible as long as the method they used complies with the PHA methodology requirement, and, more importantly, <u>all</u> facility siting hazards have been addressed (i.e., identified, evaluated and controlled). This questionnaire/checklist type of methodology would not be compliant if the site (PHA team) did not have specific justifications for each individual situation/condition that the global/generic questions addressed.*

For example, a PHA team responds "Yes" to a questionnaire/ checklist asking, "Is process equipment located near unit battery limit roads sited properly?" In this case, OSHA would first expect that the site (PHA team) would have <u>identified</u> <u>each</u> <u>location</u> where process equipment is sited near a unit battery limit road. Next, OSHA would expect the site would have <u>evaluated</u> <u>each</u> piece of process equipment located in the vicinity of a roadway. This evaluation is conducted to determine if <u>each</u> of the specific process equipment's siting is adequate/<u>controlled</u> (e.g., guarded by crash barriers, elevated on a concrete pedestal, etc.) to protect it from releasing its hazardous contents should it be *struck by vehicular traffic. Without specific justification or other specific evidence that corroborates the site's "Yes" response to this global/generic questionnaire/checklist question, a possible regulatory issue could exist for failing to address process equipment siting near roadways when it conducted its PHA.*

Guidance: Occupancy Criteria Evaluations for Employee Occupied Structure. OSHA does not accept occupancy criteria evaluations (see API 752, Section 2.5.2) as the basis for a site's determination that adequate protection has been provided for employees in occupied structures which sites have identified as being potentially subject to explosions, fires, ingress of toxic materials or high energy releases. In these occupancy criteria evaluations, the site identifies vulnerable employee occupied structures and the hazards they may be subjected to, but rather than providing protection to either the structures or employees through measures like employee relocation, spacing, or protective construction, the site simply accepts the employee exposures as adequate based on their own acceptable occupancy criteria. This occupancy criteria evaluation is solely based on the occupancy threshold criteria a site is willing to accept. For instance, API 752 list occupancy threshold criteria used by some companies as 400 personnel hours per week as acceptable exposure for employees in an occupied structure, regardless of the magnitude of the hazard these employees are potentially exposed to. The 400 personnel hours per week equates to 2 employees continually exposed in an occupied structure even if that structure has virtually no protective construction and it is sited immediately adjacent to a high pressure-high temperature reactor which contains flammable or extremely toxic materials.

<u>*Non-Essential Employees*</u>. *A site's PHA facility siting evaluation must consider the presence of non-essential personnel in occupied structures in or near covered processes. The "housing" of these non-essential employees in occupied structures near operating units may expose them to explosion, fires, toxic material, or high energy release hazards. Therefore, unlike direct support/essential personnel (e.g., operators, maintenance employees working on equipment inside a unit, field supervisors, etc.) who are needed to be located in or near operating units for logistical and response purposes, sites (PHA teams) must consider and justify why non-essential employees are required to be located in occupied structures which are vulnerable to the hazards listed above. The term "non-essential" identifies those employees who are not needed to provide direct support for operating processes. Non-essential employees include, but are not limited to, administrative personnel, laboratory employees when they are working inside a lab, maintenance*

staff when they are working inside maintenance shops/areas, and employees attending training classes.

Guidance: An example of how a temporary structure could affect a release of HHC would include a situation where a trailer's unclassified electrical system could potentially ignite flammable materials/unconfined vapor cloud if released from the process.

B. Do the PHA teams identify and evaluate all situations where operators are expected to carry out a procedure to control an upset condition, but where the operators would not have enough time to do so based on operating conditions?

C. Do the PHA team(s) identify and evaluate all situations where field employees must close isolation valves during emergencies, but where doing so would expose the employees to extremely hazardous situations? For example, to isolate a large inventory of flammable liquids, a downstream manual isolation valve would need to be closed, but the isolation valve is located in an area that could be consumed by fire.

Guidance: Some sites (PHA teams) attempt to comply with this requirement by simply addressing some global/generic human factors questions on a short questionnaire/checklist. This type of methodology would not, by itself, be adequate if the PHA team did not have specific justifications for each of its global/generic responses.

For example, if a PHA team responds "Yes" to a questionnaire/checklist asking whether emergency isolation valves (EIV) are accessible during emergencies, OSHA would then expect that the PHA team had identified, evaluated, and considered each EIV's accessibility (i.e., would the EIV be located in an area that might be consumed in fire, or is the EIV located above grade).

D. How do the PHA teams identify likely human errors and their consequences? Have appropriate measures been taken to reduce the frequency and consequences of these errors?

VIII. Operator Training

A. Have operating employees been trained on the procedures each is expected to perform?

Guidance: An "A" operator might be required to perform a different set of operating procedures than a "C" operator. Therefore, to determine if the employee has in fact been trained on the specific operating procedures they are expected to perform, cross-reference the specific procedures that an individual operator is expected to perform with the training records of the specific procedures for which the individual operator has received training. Also determine if operators perform tasks more than what is expected for their level of training.

B. From interviews with control board operators in the units, have these operators received sufficient training, initial and refresher, to be qualified to shutdown the units per the requirements of 119(f)(1)(i)(D)?

C. Based on the employer's explanation of their management of operator refresher training, verify that selected operating employees received, completed, and understood the refresher training. For each employee who operates a process, has the employer ensured that the employee understands and adheres to the current operating procedures and that the refresher training is provided at least every three years, and more often if necessary?

IX. Safe Work Practices

A. Does the site have a safe work practice which it implements for motorized equipment to enter operating units and adjacent roadways?

Guidance: "Motorized equipment" includes, but is not limited to automobiles, pickup trucks, fork lifts, cargo tank motor vehicles (CTMV), aerial lifts, welder's trucks, etc.

B. Does the site audit its safe work practices/procedures for opening process equipment, vessel entry, and the control of entrance to a facility or covered process area?

C. Does the site have a safe work practice for opening process equipment, e.g. piping and vessels, and does the site require their employees and contractor employees to follow it?

X. Incident Investigation Reports

A. Provide a list of actual incidents and near-miss incidents that occurred at the site within the last year. Have all factors that contributed to each of the incidents been reported and investigated?

Guidance: An "actual incident" is defined as an incident with negative consequences such as a large HHC release, employee injuries or fatality, or a large amount of property or equipment damage. Typically, based on loss-control history, there is a much higher ratio of near-miss incidents in the chemical processing and refining industries than there are actual incidents.

XI. Blowdown Drums and Vents Stacks (Blowdowns)

A. Does the site have any blowdowns? If so, does the PSI include the original design and design basis for each blowdown at the site?

Guidance: Blowdown(s)—refers to a piece of disposal equipment in a pressure-relieving system whose construction consists of a drum to collect liquids that are separated ("knockout") from vapors and a vent stack, which is an elevated vertical termination discharging vapors into the atmosphere without combustion or conversion of the relieved fluid. Blowdown(s) are separate vessels intended to receive episodic (e.g., when de-inventorying a vessel for a planned shutdown) or emergency discharges. Blowdown(s) are designed to collect liquids and to dispose of vapors safely. In the refinery industry, hydrocarbons typically enter blowdown(s) as liquids, vapors, or vapors entrained with liquids. Blowdown(s) typically include quench fluid systems which reduce the temperature of hot, condensable hydrocarbons entering the blowdown as well as the amount of vapor released via the vent stack. These systems can include internal baffles to help disengage liquids from hydrocarbon vapors. Sometimes, blowdown(s) include inert gas or steam systems to control flashback hazards and to snuff vent stack fires if ignited by sources such as lightning.

Examples of PSI related to blowdowns, their design and design basis include, but are not limited to, such items as:

1. *Physical and chemical properties of the materials relieved to blowdowns (See API STD 521, Section 6.2.1);*

Guidance: Of particular concern are heavier-than-air hydrocarbons with relatively lower boiling points. Additionally, hot hydrocarbons pose a greater risk because they are more volatile. Releasing these materials under the right conditions can result in the formation of unconfined vapor clouds which can and have resulted in major catastrophes at refineries and chemical plants.

2. *A definition of the loadings to be handled (See API STD 521, Section 7.1);*

3. *The exit velocity of gasses/vapors released from the vent stack (See API STD 521, Section 7.3.4);*

4. *Design basis/"worst-case" scenario for maximum liquid—vapor release to blowdown (See API STD 521, Section 4.5.j and 7.1.3);*

5. *When more than one relief device or depressuring valve discharges to a blowdown, the geographic locations of those devices and valves must be defined (See API STD 521, Section 4.4.q. and 7.2.3);*

6. *The design residence time of vapor and liquid in the drum (See API STD 521, Section 7.3.2.1.2);*

7. *The design basis for the vapor—liquid separation for the drum;*

8. *The design basis for the exit velocities for the vent stack; and*

9. *The nature of other, lesser hazards related to smaller releases not related to the design "worst-case" scenario such as the release of toxic (e.g., H_2S) and corrosive chemicals.*

B. Since the original installation of the blowdowns, have the original design and design basis conditions remained the same? If not, was an MOC conducted to determine if the blowdown design and capacity are still adequate?

Guidance: Examples of conditions that may have changed since the original design and installation of the blowdowns include: increased throughput in the unit(s) that relieve to the blowdowns; additional relief streams routed to the blowdown, blowdowns originally designed only to handle lighter-than-air vapor emissions from their stacks have had liquids or other heavier-than-air releases emitted from their vent stacks; additional equipment, a new unit, or occupied structures have been sited near the blowdowns in a manner that was not addressed in the original design or design basis, etc.

C. Did the PHA identify all scenarios where hot, heavier-than-air, or liquid hydrocarbons might be discharged from blowdown stacks to the atmosphere?

D. Can the site demonstrate that atmospheric discharges from blowdowns are to safe locations?

Guidance: Other structures such as control rooms, trailers, offices, motor control centers, etc., must be considered in a PHA to determine if they have been sited in a safe location that might be affected by a hydrocarbon or toxic material release from a blowdown. Unsafe locations can include, but are not limited to, the location of equipment which could act as an ignition source, such as a furnace stack; an employee platform on a column where employees would be exposed in the event of a release; a control room; a satellite building; a trailer; a maintenance area/shop; an emergency response building; an administration building; a lunch or break room; etc.

E. If there is a high-level alarm in the blowdown drum, is there an MI procedure for calibrating, inspecting, testing and maintaining the instrument/control?

Guidance: The required documentation data must include the date of the inspection or test, the name of the person who performed the inspection or test, the serial number or other identifier of the equipment on which the inspection or test was performed, a description of the inspection or test performed, and the results of the inspection or test.

F. Have blowdown operators received appropriate training, either initial or refresher?

7

SAFETY AND HEALTH TRAINING

WHY TRAIN?

> Ensure that all employees understand the hazards to which they may be exposed and how to prevent harm to themselves and others from exposure to these hazards, so that employees accept and follow established safety and health protections.[1]

> Training is necessary to reinforce and complement management's commitment to prevent exposure to hazards. All employees must understand the hazards to which they may be exposed and how to prevent harm to themselves and others from such hazard exposure.[2]

> The employer shall instruct each employee in the recognition and avoidance of unsafe conditions and the regulations applicable to his work environment to control or eliminate any hazards or other exposure to illness or injury.[3]

These statements emphasize the importance that OSHA places on safety and health training for all employees. That significance gains more emphasis when one reviews the OSHA Fatal Facts on the OSHA website. Of the 71 listed facts on the website, 49 had inadequate training as a contributing factor to the accident and fatality.[4] The underlying principle is that all those with responsibilities to perform specific acts must receive adequate training to be able to perform those acts properly and in a safe and healthful manner. This principle is relevant during all stages and activities of our lives. When our parents wanted us to progress from the diaper to the potty and from the bottle to the sippy cup to the glass, when they bought us our first bicycle,

Preparing for OSHA's Voluntary Protection Programs. By Brian T. Bennett and Norman R. Deitch
Copyright © 2010 John Wiley & Sons, Inc.

when we got our first cars, when we began to work, we were more successful when we received training that focused on the expectations and the proper way to perform those tasks. How many of us have had to clean up after a child that had an accident, fell off a bike, got that first ding in our cars, and got hurt at work? Without adequate training in the workplace we can expect accidents and injuries to occur with more frequency than with adequate training. Recognizing this, OSHA has placed a great deal of emphasis on safety and health training. Many OSHA standards contain a specific detailed requirement for employee safety and health training. Some of those OSHA standards include:

- Employee Emergency Action Plans, 29 CFR 1910.38
- Process Safety Management, 29 CFR 1910.119
- HAZWOPER, 29 CFR 1910.120
- Respiratory Protection, 29 CFR 1910.134
- Permit Required Confined Space, 29 CFR 1910.146
- Control of Hazardous Energy, 29 CFR 1910.147
- Powered Industrial Trucks, 29 CFR 1910.178
- Bloodborne Pathogens, 29 CFR 1910.1030
- Hazard Communication, 29 CFR 1910.1200
- Construction Safety Training and Education, 29 CFR 1926.21
- Woodworking Tools, 29 CFR 1926.304
- Scaffolding, 29 CFR 1926.404
- Fall Protection, 29 CFR 1926.503
- Excavations, Trenching and Shoring, 29 CFR 1926.651

Many other OSHA standards also specifically require employee training. These can be found in OSHA publication 2254, Revised 1998, Training Requirements in OSHA Standards and Training Guidelines.

VPP TRAINING REQUIREMENTS

The VPP process focuses on training for managers, supervisors, production employees, contractors, and all others that may be exposed to the hazards of the workplace. Since training must be relevant to the responsibilities of the trainee, not all training is applicable to all groups. Therefore it is critical that development of training programs take into consideration the jobs and tasks of the employee and the hazards to which the employees may be exposed.

TRAINING NEEDS ASSESSMENT

Development of training programs should start with a training needs assessment based on position and job requirements and the results of the work-site analyses as discussed earlier. The training needs must also include that training that is required by OSHA standards, company procedures and programs, and performance expectations. Resources that can be reviewed for the training needs assessment may include:

- Job hazard analyses
- Task hazard analyses
- Performance standards
- Injury and illness trends at the workplace
- Incident reports
- Injury and illness trends that have been recognized at similar work sites
- Findings from routine inspections
- Safety maintenance work orders
- Recommendations for improvement from safety and health evaluations
- Current trends and emphasis programs identified by OSHA
- Employee reports of unsafe conditions or practices

To foster employee involvement and obtain additional input, the training needs assessment should provide for a means for affected employees to be involved. For example, it would be prudent to include a member of the confined-space rescue team to explain the training needs of the team. Other examples include involving mechanics and chemical operators in assisting in the development of the training needs assessment for control of hazardous energy and process safety management. Since the employees are directly associated with the operation, they are in an excellent position to recommend training. It must be recognized, however, that the employees may not be those most familiar with the training requirements of the applicable standards. Therefore, the training needs assessment process must also involve specific subject matter and standards experts or resources.

The training needs assessment must consider the responsibilities of all personnel at the work site, including administrative, temporary, and contractor employees. The assessment must also consider the performance requirements of all employees so that they can be provided with the knowledge and skills to meet those requirements.

DEVELOPMENT OF TRAINING PROGRAMS

Once the training needs assessment is complete, specific training programs should be developed to meet the identified training needs. These training programs may be developed in house by managers, supervisors, and employees. Off-the-shelf and tailored training programs may also be purchased from outside training development companies. Another source for training programs would be to use the services of an outside organization to provide nationally recognized training programs such as the American Heart Association for the cardiopulmonary resuscitation (CPR) training program or the National Safety Council for the fork truck training program.

The training needs assessment should also help to develop an annual training plan. The training plan is a tool used to schedule training by subject and either employee or position. The plan would identify when each training program will be delivered and what positions must receive what training. For example, most administrative staff would not be required to use PPE. However, if an employee in the administrative position is required to visit the production areas, he or she should receive the same PPE training as is required for all employees in those areas. Development of the annual training plan also helps to facilitate the tracking of training.

As evidence of the completion of training, records must be maintained. The usual method is the use of sign-in sheets. A more effective method is to maintain the tests for each class. There was at least one workplace where the employees refused to sign in for training. Their theory was that if they did not acknowledge the training, they could not be held responsible for violations of procedures or rules. The management correctly listed all those in attendance and verified the list by signing it. Another very effective method to track training is through the use of computer-based training that is linked to the tracking system.

TRAINING THE ADULT LEARNER

Since the training will be provided to the employees and is critical for their safety and health, the training programs must be carefully designed and implemented to ensure that they will be successful. There are many theories on effectively training the adult learner. The underlying principle is based on how we learn and what we remember from that learning. Based on a standard learning pyramid, generally, the adult learners remember only about 5–10% of what they are told and read, about 30% of what is actually demonstrated, and about 75% of what they practice. The most effective method of reinforcing training (about 90% retention) is to teach others and to use the lessons taught immediately. Figure 7.1 is the Learning Table from the National Training Laboratories, Bethel, Maine. Considering this, it is critical

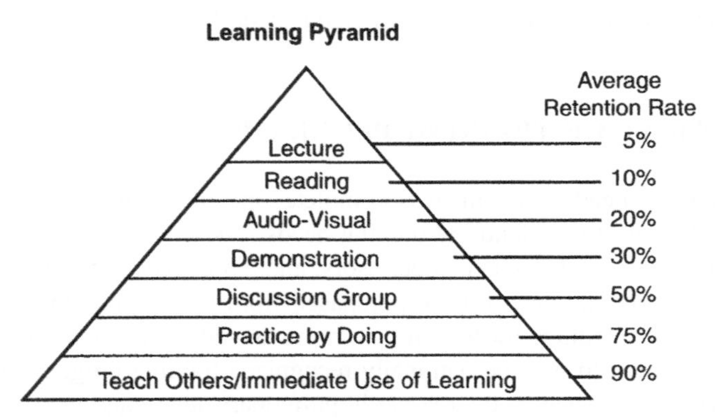

Figure 7.1 Learning Table from the National Training Laboratories, Bethel, Maine.

that safety and health training emphasize training on how to perform the expected tasks using techniques that require active employee participation. Employees should also be encouraged with rewards to provide training to their co-workers. That is not to minimize the use of lecture and demonstration tools. These are also valuable tools to explain the information and to demonstrate how the task is to be performed.

Another critical factor in training the worker is to ensure that the material presented is understood and that the employees are able to demonstrate those skills as expected. Although most OSHA-required training does not include a requirement for verification of the knowledge and skills of the trained employees, there are several that do. One such example is the Powered Industrial Truck standard, 29 CFR 1910.178. This standard has a requirement to perform a written evaluation of the effectiveness of that training to ensure that the operator has the knowledge and skills needed to operate the powered industrial truck safely.

VPP TRAINING EXPECTATIONS

The safety and health training element of the VPP contains several subelements that must be addressed in all VPP activities including the application, the annual evaluation, and the OSHA VPP onsite evaluation. These subelements include:

1. Managers and supervisors understand their safety and health responsibilities and are able to carry them out effectively.
2. Managers, supervisors, and nonsupervisory employees (including contract employees) are made aware of hazards and are taught how to recognize hazardous conditions and the signs and symptoms of workplace-related illnesses.
3. Managers, supervisors, and nonsupervisory employees (including contractor employees) learn the safe work procedures to follow in order to protect themselves from hazards, through training provided at the same time they are taught to do a job and through reinforcement.
4. Managers, supervisors, nonsupervisory employees (including contractor employees), and visitors on the site understand what to do in emergency situations.
5. Where personal protective equipment is required, employees understand that it is required, why it is required, its limitations, how to use it, and how to maintain it; and employees use it properly.

In addition to evaluating each of these subelements, each training program should be routinely evaluated to ensure that they are effectively meeting the needs of the employees to enable them to work safely. The training programs must also be evaluated to ensure that they are current.

Evaluation of the subelements will be addressed first, with a discussion of evaluating the training programs after. Finally, we will discuss the benefits and issues related to each training method.

EVALUATION OF THE TRAINING ELEMENTS

For all training, the following must be considered:

1. *Training records must be maintained*: OSHA will review the training records to confirm that training has been delivered and to determine how the VPP work site verifies that all required training is completed.
2. *Training must be appropriate to the work performed and the hazards present*: OSHA will review selected training curricula to determine that the material is correct, complete, and meets the requirements of the work site. The selected curricula should include the training that is required by OSHA standards and the training that addresses the more serious hazards at the work site.
3. *Training programs are evaluated for effectiveness*: Relying primarily on employee interviews and observations of work performance, OSHA will determine if the training has been effective. Employees should be able to explain work and emergency procedures and the proper use of PPE. They should also demonstrate safe operational behaviors.
4. *There is an annual training plan that facilitates the tracking of all training*: OSHA will review the annual training plan to determine if it is appropriate to the needs of the employees. They will also inquire about how the training plan was developed and if employees were involved.
5. Employees are tested to ensure that they have learned the material and skills presented; test results will be reviewed.

Each of the VPP-identified elements may be evaluated using the following tools:

1. Managers and supervisors understand their safety and health responsibilities and are able to carry them out effectively. The following should be reviewed:
   a. Use the training tracking reports to confirm that all managers and supervisors have received all required training identified in the annual training plan.
   b. Review the attendance lists at routine safety meetings to confirm attendance by managers and supervisors.
   c. Observe managers demonstrating the skills taught in the training.
   d. Comments from direct reports regarding the compliance of managers with the training material.
   e. Interview managers and supervisors to discuss the training material and to confirm that they know their responsibilities.
2. Managers, supervisors, and nonsupervisory employees (including contract employees) are made aware of hazards and are taught how to recognize hazardous conditions and the signs and symptoms of workplace-related illnesses. The following should be reviewed:
   a. Review the annual training plan to determine if it includes hazard recognition training for all employees (including managers and supervisors).

b. Interview employees (including managers and supervisors) to determine their awareness of the hazards present in the workplace.
c. Review the contractor orientation and training program to ensure that it explains the hazards at the workplace.
d. Review the training records to determine that all contractor employees have received the contractor orientation.
e. Interview work-site employees (including managers and supervisors) and contractor employees to determine their ability to recognize hazards.
f. Interview work-site employees (including managers and supervisors) and contractor employees to determine their awareness of the hazards to which they may be exposed.
g. Interview work-site employees (including managers and supervisors) and contractor employees to determine their awareness of the safe work procedures.
h. Review inspection reports to identify those hazards that are found with a frequency that is greater than would be expected if hazard recognition training was effective.
i. Review personal performance evaluations of managers and supervisors to determine how they are being held to the training requirements.

3. Managers, supervisors, and nonsupervisory employees (including contractor employees) learn the safe work procedures to follow in order to protect themselves from hazards, through training provided at the same time they are taught to do a job and through reinforcement. The following should be reviewed:
   a. Review the training records to determine that all employees have received the required training.
   b. Observe all workers periodically to determine if they are performing the tasks as they were trained.
   c. Interview randomly selected employees to determine their knowledge of the training provided.
   d. Review all disciplinary actions to determine if failure to work as trained has resulted in any disciplinary actions for management or production employees and contractors.
   e. Review the tests and test results for all training.

4. Managers, supervisors, nonsupervisory employees (including contractor employees), and visitors on the site understand what to do in emergency situations. The following should be reviewed:
   a. Review the training records to determine that all employees have received emergency action training.
   b. Review the critiques from previous emergency action plan drills, such as evacuations.
   c. Interview employees to determine their knowledge of the emergency action plan.

5. Where personal protective equipment is required, employees understand that it is required, why it is required, its limitations, how to use it, and how to maintain it; and employees use it properly. The following should be reviewed:
   a. Observe employee use of PPE in the workplace.
   b. Interview employees to determine if they understand the need for the PPE.
   c. Interview employees to determine if provided PPE is suitable for them.
   d. Observe the conditions of the PPE employees are using to determine their ability to properly maintain it.
   e. Observe how employees are cleaning and storing their PPE to determine if they are doing it as they were trained.

TRAINING FOR OFF-THE-JOB ACTIVITIES

Clearly, the previous discussion was focused on training related to the work and hazards at the workplace. Since the majority of employees sustain serious injuries while not at the workplace, consideration should also be given to providing training to all managers and employees that will enable them to avoid non-work-related injuries and illnesses. This type of training is becoming very prevalent in workplaces in the United States and is usually referred to as the 24/7 program to indicate that the focus is to protect the employee at all times. While providing the employee with additional training, the company benefits as a result of the reduction in off-work injuries that prevent employees from returning to work or may become exacerbated at work. This training is intended to provide the employees the tools and incentives to use the safety knowledge they receive at work to their nonwork activities. Typical subjects for this training include working at heights to avoid standing on chairs, cooking to avoid burns and food poisoning, driving and all of the hazards associated with it and the focus on seatbelt use and distracted or impaired driving, smoking cessation, wellness, camping and other recreational activities and exposure to the elements, hand washing to limit the spread of infection, home emergency action plans, chemical safety in the home, and the use of personal protective equipment at home. Several companies provide their employees with extra personal protective equipment such as hearing and eye protection and gloves to protect employees while they work at home. Others provide their employees with either smoke and fire detectors or batteries for them.

TRAINING METHODS

Training can be developed using many techniques. Each has its own benefits and issues. These include:

- Lecture
- Video
- PowerPoint

- Demonstration
- Computer based
- Open discussion
- Role play
- Field trips

All training programs should have stated learning objectives for the class. These learning objectives should be reasonable, specific, directly related to the material, and measureable. Many learning objectives do not meet these requirements. The biggest error in developing a learning objective is to focus on the act of teaching. For example, a typical learning objective may be: "This class will explain the principles of the Voluntary Protection Programs (VPP)." Although the presenter may be successful in reaching that goal by simply discussing the VPP, the learners may not understand the information. Instead of a learning objective, this is actually a teaching objective. A better solution would be to list specific goals of the class, such as:

At the conclusion of this class the learners will be able to:

- Explain the principles of the VPP.
- List at least four benefits of the VPP.
- Identify the four major elements of a VPP-type safety and health management system.
- Describe the application process.
- Explain the Special Government Employee Program and the Mentoring Program.

The presenter must also be cognizant of the learning needs of the class. Are they there simply to meet the training requirement of the standard or the company, or are they there because they have a strong interest in the subject. A third or more frequent repetition of the annual hazard communication training will not be received in the same way as a CPR class for new volunteers. Even though a subject may have been covered in the past, it may have a new importance because of recent occurrences. Emergency action plan training and evacuation drills gained a new emphasis after September 11, 2001, and preparation for hurricanes gained in importance after the devastation wrought by Hurricane Katrina.

The learning objectives, length of the class, material, and presentation may have to be adjusted based on the learning needs of the class. Those that are there because they have a strong interest in the subject may need less emphasis on the importance of the material and more detail about the subject. Those that are new to the subject would probably need a greater explanation of the subject and its importance instead of more detail. The detail may be provided in a later session.

LECTURE-BASED TRAINING

Given that the learning pyramid indicates that adult learners only retain about 5–10% of what they are told or read, it is clear that that is the least effective method of training. It is unfortunate then that lecture and reading are the most common methods of training. Many construction projects provide regular weekly training to the workers at the project. That training generally is given by the trades' superintendent by reading a one-page document received from a subscription service. Copies of those documents may also be given to each of the workers. To compound this problem, many of these training sessions, commonly referred to as "toolbox talks," are repeated each year so that they become more routine and hence boring to the listener. To compound the problem even more, the reader may not be technically competent to deliver the safety message and just reads the message without any explanation. These types of sessions are not limited to the construction industry and many general industry work sites use them as well. Fortunately, although still very common, this type of training is slowly being phased out.

That is not to say that all lectures are ineffective. Many such presentations can be very successful depending on the interest of the subject to the learner and the skill of the instructor. In fact most training involves some element of lecture, but it is the level of the quality of the lecture that is the concern with this method.

AUDIO VISUAL MATERIALS

The next most effective method of training is the use of audio visual materials. These methods are very effective in getting the attention of the learners and maintaining their interest. For example, when discussing electrical safety, the learner's attention will be captured by presenting a video of an arc flash blast. There are many such videos available for many safety and health subjects. They may be either humorous or very serious, but they are very effective attention getters. Many commercial companies develop and sell training videos on numerous subjects. These are useful for many subjects that are fairly generic such as bloodborne pathogens or first aid and CPR. However, if the subject matter is very technical and site specific, such as confined-space entry or control of hazardous energy, they must be supplemented with the site-specific procedures and rules. Even the bloodborne pathogens training must refer to the site exposure control plan. One problem that has been noted with these videos is that the employees have seen them many times over the years and lose their interest in them.

SLIDE PRESENTATIONS

The next, and probably the most widely used method, is PowerPoint slide programs. Again, these may be generic off-the-shelf or homemade programs. Although the homemade programs may not have the professional touch of the commercial products, they may actually be more effective. Many that have been reviewed by the authors during numerous OSHA VPP onsite evaluations have been excellent. They make up for the lack of professionalism of the commercial products with their imagination and focus on the programs and procedures of the work site. Many of them have been developed by hourly employees who then obtain a greater sense of ownership of the subject. This method of developing training materials also results in the reinforcement of the subject matter knowledge of those directly involved. They may also assist in the presentation of the training.

When using PowerPoint slides, there are several important guidelines or rules that should be observed. As discussed earlier, a presenter should avoid reading directly from notes. The learner will most likely think that they could have learned the same material by reading the notes themselves. The presenter should use the notes as talking points or reminders of the specific material that is being covered. Therefore, PowerPoint slides should consist of talking points or an outline of the material. One rule of thumb is that a standard slide should not have more than five lines and that no line should have more than seven words. That is not an absolute rule, but it illustrates that the information on the slide is not a comprehensive explanation of the material but rather a group of bullets or talking points. The presenter should explain each bullet in detail and the learner should take appropriate notes.

The bullet slides can be supported with more detailed slides, but only as examples. For example, when discussing a permit required confined-space program, it may be appropriate to include a slide that contains the OSHA chart that describes the difference between a permit and a nonpermit confined space. Another example of an exception may be to include on a slide a copy of the work-site's safety and health policy. The purpose of including such a slide would be to allow a discussion of the intent of the policy and where it is located.

A third exception may be to include slides with pictures or news articles to emphasize points that the presenter is trying to make. The most effective use of pictures is to illustrate real workplace conditions or to focus on the consequences of disregarding the safety and health rules. Pictures are great attention getters and help to break up the lecture and maintain interest. They also are usually more effective than a verbal explanation of what something is supposed to look like. Consider the old saying that "a picture is worth a thousand words."

Figure 7.2 illustrates an acceptable PowerPoint slide. Figure 7.3 illustrates a poor PowerPoint slide because it is too wordy. Although the slide in Figure 7.3 is too wordy given the criteria previously discussed, it can be used to simply illustrate the policy. It is not intended to be read. It is actually a copy of the first eight of the required VPP assurances.

Figure 7.4 illustrates a slide to lead into a discussion of the OSHA Confined-Space Decision Tool. That is used to demonstrate the OSHA chart so that the learners may become familiar with it. The slide should be supported with the actual chart handed to each learner.

Similar to the slide of the policy, this illustration can be discussed as a tool that is available to identify a confined space. Figure 7.5 is an example of a picture to lead to a discussion of electrical safety, provided by EHS Excellence Consulting, Inc.

A critical factor in most effective training is to encourage audience participation. There are several methods that can be used to involve the audience in the presentation. The easiest is to simply ask the audience members to discuss their specific experiences related to the subject. For example, when discussing ground fault circuit interrupter (GFCI) devices, the audience can be asked to identify those articles in their homes

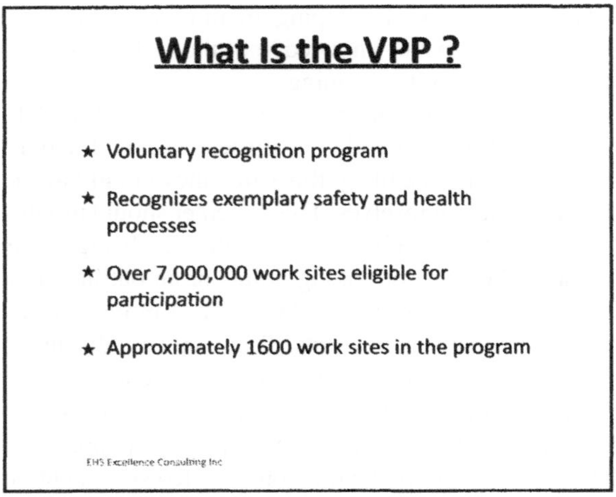

What Is the VPP ?

- ★ Voluntary recognition program
- ★ Recognizes exemplary safety and health processes
- ★ Over 7,000,000 work sites eligible for participation
- ★ Approximately 1600 work sites in the program

EHS Excellence Consulting Inc

Figure 7.2 Example of an acceptable PowerPoint slide.

that have a built-in GFCI (hair dryers). Role playing is also a very effective tool to involve many people. It also can lead to some humorous situations. When discussing lockout/tagout, have two or three people simulate the process. This can be supported by requiring them to actually apply some locks and use a lockbox while playing the roles of authorized maintenance and affected operational personnel. They can also play the role of the supervisor who is responsible for ensuring that the procedures are performed correctly. A wrinkle could be to have the "supervisor" instruct the "authorized employee" to disregard a step.

Yet another example of role playing can be used to illustrate the enforcement of the safety and health rules. This lends itself very easily to a role reversal wherein the hourly employee acting as the supervisor counsels the offending supervisor for not wearing the hearing protection. That counsel should include the reason and importance of the hearing protection and the proper way to use it.

Another critical factor in any training is the actual delivery of the presentation. Regardless of the tools used, whether they are demonstration, slides, movies, role playing, or any other method of training, it is ultimately the lead and other instructors or presenters that will set the tone of the training session. Most, if not all of us, have sat through the endless monotone droning of a lecturer who reads directly from notes. We have also probably sat through a session wherein the lecturer had a habit of uttering a verbal pause (e.g., um, uh, mm) after every thought. Another habit that can have a detrimental effect on a training session is the presenter that exhibits a total lack of interest in the subject. The presenter may speak in a low voice, look down or away, disregard student questions, constantly look at his or her watch, or in any other way demonstrate that he or she would rather be somewhere else. Another way to turn off the students is to exhibit a strongly negative body language, such as standing firm with arms crossed in front or hands deep in the pockets of the pants. One more negative presentation trait that is an interest turn off is the attitude

VPP Assurances

As part of this application, () is proud to submit these VPP assurances:

1. Compliance

() will comply with the*Occupational Safety and Health Act (OSH Act*) and correct in a timely manner
all hazards discovered through self-inspections, employee notification, accident investigations, OSHA onsite reviews, process
hazard reviews, annual evaluations, or any other means. () will provide effective interim protection, as necessary.

2. Correction of Deficiencies

Within 30 days, () will correct safety and health deficiencies related to compliance with OSHA
requirements and identified during any OSHA onsite review.

3. Employee Support

() employees support the VPP application. At sites with employees organized into one or more
collective bargaining units, the authorized representatives for each collective bargaining unit for the majority of the covered e
mployeeshas either signed the application or submitted a signed statement indicating that the collective bargaining agent(s)
support VPP participation. At non-union sites, management's assurance of employee support will be verified by the OSHA
onsite review team during employee interviews.

4. VPP Elements

VPP elements are in place, and management commits to meeting and maintaining the requirements of
the elements and the overall VPP.

5. Orientation

Employees, including newly hired employees and contract employees, will receive orientation on the
VPP, including employee rights under VPP and under the *OSH Act.*

6. Non-Discrimination

() will protect employees given safety and health duties as part of () safety and health
program from discriminatory actions resulting from their carrying out such duties, just as Section 11(c)
of the*OSH Act* protect employees who exercise their rights.

7. Employee Access

Employees will have access to the results of self-inspections, accident investigations, and other safety
and health data upon request. At unionized construction sites, this requirement may be met through
employee representative access to these results.

8. Documentation

() will maintain our safety and health program information and make it available for OSHA
review to determine initial and continued approval to the VPP. This information will include:

•Any agreements between management and the collective bargaining agent(s) concerning safety
and health.

•All documentation enumerated under Section III.J.4. of the July 24, 2000*Federal Register* Notice.

•Any data necessary to evaluate the achievement of individual Merit or 1-Year Conditional Star goals.

Figure 7.3 Example of a poor PowerPoint slide with too much information.

that the presenter exhibits to the audience. Sloppiness, slovenliness, unkempt appearance, or bad odor are traits that imply that the owner has no interest in personal appearance, which can lead one in the audience to suspect that the presenter has no interest in anything else, such as the subject matter being presented. The attitude of the presenter has a direct effect on the audience. The material will be perceived as not important enough to pay attention and the primary interest will be focused on when it will be over.

Given the above observations, it is clear that a training presentation will be enhanced by a presenter that exhibits a positive appearance and who can extend that positive attitude to the material. It is possible for an excellent presenter to make

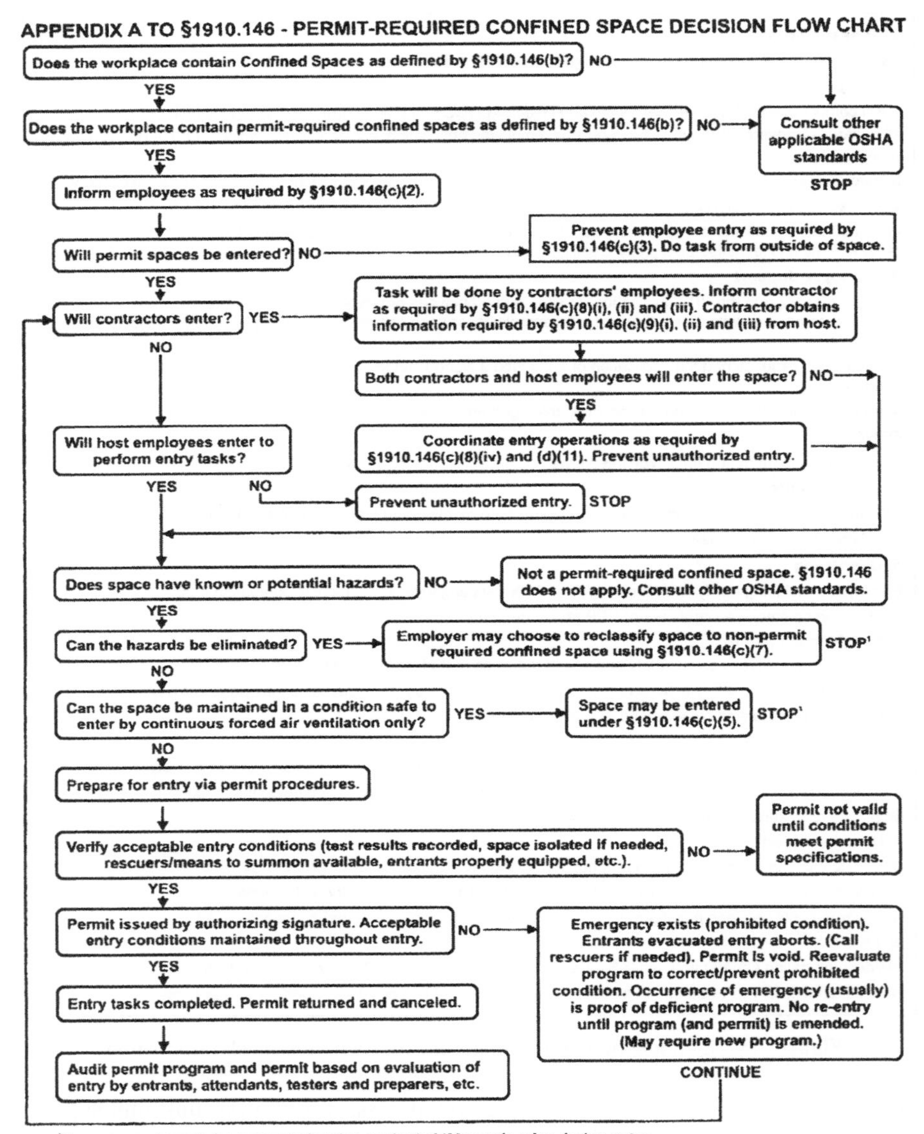

Figure 7.4 Example of an illustration supported with a handout.

even the most mundane and typically boring information seem interesting and important. It is also possible for a less capable presenter to take an interesting subject and make it boring.

Keys to a successful presentation mode include:

- Neat appearance
- Clear speaking voice

Figure 7.5 Picture used to discuss electrical safety (provided by EHS Excellence Consulting, Inc).

- Modulated speaking
- Direct eye contact
- Opportunity to ask questions
- Allowing questioner to complete the thought
- Allow an adequate response time when asking for input or questions
- Use leading questions to generate discussion
- Responsive to questions
- Direct one-on-one interaction
- Open posture

On the opposite side of the equation, the following are traits to avoid:

- Direct reading of material
- Ignoring or refusing to answer questions
- Not allowing questions to be completed
- Indicating any disrespect toward the audience
- Hiding behind tables, chairs, or other physical barriers such as a podium

Posture is critical for successful presentations. The ideal position for a lecturer is to stand at what is known as the "ready position." The ready position is when the lecturer is standing facing the audience with arms open at shoulder width and bent at the elbow so that the lower arm is parallel with the ground. This allows the lecturer to freely use the arms to point at a screen, flip chart, or another person. It also allows for free use of hands and arms to emphasize significant points or to address specific audience members.

Presenters should be encouraged to use notes to ensure that they identify and include all relevant points of the subject matter. These are used as talking points. In fact, that is the key to a successful presentation. Rather that reading or speaking to the audience, the successful presenter talks to and with the audience. The talking points should be organized to maintain a smooth flow of the information. Discussing the use of a lockbox before discussing the lockout procedure could confuse the audience.

The timing of a training program is also critical to its success. It has to be long enough to adequately cover the material, but not too long that the learners will lose interest. The OSHA Training Institute has found that a session duration of 50 minutes with a 10-minute break has been very successful. It important that the class follow whatever schedule has been agreed to. It is also critical that all of the material must be presented within a specific agreed amount of time.

Many successful training programs have failed because of situations over which the instructor may have no control. Some examples of what may lead to the failure of what would otherwise be a successful program include:

- Inadequate room size
- Poor room lighting
- Inadequate ventilation
- Inadequate temperature control
- Poor acoustics
- Malfunctioning equipment

Before the class begins, the presenter should inspect the classroom to ensure that it meets the needs of the presenter and the learner. The above concerns should be inspected and addressed before the class begins. If using PowerPoint or video, the presenter should have backup equipment such as a replacement projector or at least a spare projector bulb. The chair and desk arrangement may be adjusted to make it more suitable for the type of presentation. If planning on group work, round tables are usually more suitable than rectangular ones. Walking around the class may be impossible if the setup is podium based and there is no remote control to advance slides. These are some factors that must be considered.

The best way to determine the success of the training is to ask the learners. One measure of success is their successful completion of a test or some other form of knowledge verification. The authors strongly support the idea that all training should end with some form of knowledge verification, even if it is only a measure of classroom participation. That is the best way to gauge the effectiveness of the material and the presentation. Another more formal method is to have each learner complete an evaluation of the class.

The evaluation should address at least the following areas:

- Relevance of the subject
- Relevance of the material

- Usefulness of the material
- Effectiveness of the presenter
- Effectiveness of the presentation
- Description of how the material will be used by the learner
- Suggestions for additional material to be discussed
- Suggestions for material that should be deleted
- Identifying the strongest and weakest parts of the presentation
- Any and all suggestions to improve the program

The most important part of any class evaluation is to heed the suggestions that are provided. They are usually valid, and failure to heed them may result in the failure of future classes. The word will get around that the presenter is inflexible and does not listen to reason. That will lead to a negative attitude toward the class to start, and it will be an uphill battle from there.

REFERENCES

1. OSHA, Safety and Health Program Guidelines, January 26, 1989.
2. *Federal Register*, Vol. 74, No. 6, 1/9/09; revisions to the VPP, IV. D.
3. 29 CFR 1926.21(b)(2).
4. http://www.osha.gov/OshDoc/toc_FatalFacts.html.

8

THE OSHA ONSITE VPP EVALUATION

APPLICATION SUBMITTAL

Now that the application has been written and reviewed, it is ready to be submitted to the OSHA VPP Regional Manager. Although there will probably be a tendency to submit the application to OSHA in draft form, that is not necessary! The first thing OSHA will do upon receipt of the application will be to acknowledge that the application has been received, usually by e-mail. That acknowledgment must be made within 15 days of receipt. Someone in OSHA will then review the application to ensure that all of the questions have been answered in sufficient detail to explain how each of the safety and health management system elements and subelements are managed. The review will also ensure that all necessary attachments are provided. OSHA will also confirm the calculations on the tables of injuries and illnesses. They will also review the inspection history of the work site to ensure that there are no open citations and no upheld willful violations issued in the last 3 years.

Once OSHA has reviewed the application, they will notify the work-site VPP coordinator that the application has either been accepted or that there are some questions that have to be explained. The questions are usually along the lines of providing additional information or detail, or providing supporting examples of how you manage certain aspects of the system. One example of a request for more detail would be to ask the applicant to explain how he or she assures that all areas of the work site are inspected at least quarterly. A valid response may be that the plant is sectioned into three areas and each area is thoroughly inspected on a set schedule each month.

Preparing for OSHA's Voluntary Protection Programs. By Brian T. Bennett and Norman R. Deitch
Copyright © 2010 John Wiley & Sons, Inc.

The inspections and results are then documented in some form of electronic tracking system. Another example is to explain the detail of the staff performance evaluation process, including a copy of a template or sanitized performance evaluation form.

The applicant then has 90 days to respond to OSHA with the requested information and attachments. Of course, you can always contact OSHA to ask that they explain their requests. You can also contact them to explain anything that you either cannot, or would rather not, provide. If, for example, your performance evaluation system is proprietary and should not be included in the application, OSHA should be able to accept just a detailed description of the process with the agreement that the documents will be reviewed during the OSHA VPP onsite evaluation.

If you need more than 90 days to provide the requested information, you must ask OSHA to allow you additional time. Although not required to do so, OSHA has traditionally taken the position to be liberal in responding to such requests. If you do not provide the requested information and do not request an extension to file, OSHA is required to return the application. Such an action does not officially preclude a resubmission in the future, but it may not be well received by OSHA. That is something that should be addressed with the local OSHA VPP Regional Manager.

Congratulations! OSHA has now reviewed your application and additional submissions and officially notified you that the VPP application has been accepted. Now what?

OSHA will contact the worksite VPP coordinator to schedule the OSHA VPP onsite evaluation (the Onsite). The Onsite is scheduled at the convenience of all participants but must be completed within 6 months of the acceptance of the application. OSHA will schedule the Onsite to observe your normal operations; they try to avoid special activities such as maintenance outages or holiday reductions in operations. When you are contacted by OSHA to schedule the Onsite, consider your planned activities and the schedules of your management and safety and health staff.

During the Onsite, OSHA will interview the management staff and will work closely with the safety and health staff. Therefore, you must also consider their scheduled absences for vacation, meetings, conferences, or anything else. One of the authors scheduled an Onsite at a very large manufacturing plant in upstate New York in early November. When the Onsite team arrived, they found out that about 50% of the staff was absent because that was the first day of deer hunting season. This is one example of those things to consider when scheduling the Onsite evaluation.

The schedule for the Onsite is critical since it provides you with a known window of opportunity to prepare for the Onsite team. It also identifies that date 75 days earlier that your work site becomes exempt from routine OSHA programmed compliance inspections. It also provides OSHA the time to prepare their onsite team and review relevant documents for the upcoming evaluation.

PREPARING FOR THE VPP ONSITE EVALUATION

Now we will discuss the preparation for the Onsite, first from OSHA's perspective and then from the applicant's. We will address OSHA first because their needs will lead to how the work site prepares.

OSHA will first form a VPP onsite evaluation team (Team). The Team composition will be based primarily on the size and complexity of the workplace. The Team will be composed of from three to five or more individuals with various backgrounds. There will always be an OSHA Team leader, with additional Team members that will include at least one safety specialist and one industrial hygienist. If there are significant complex hazards at the workplace, the Team will have a specialist in that field. For example, if there are chemicals present that are covered by OSHA's Process Safety Management Standard (29 CFR 1910.119), the Team will have at least one specialist who can review the related activities in detail. Another example would be to use a crane specialist for construction work sites.

Many Teams are composed of OSHA special government employees (SGE). These are individuals that work for other VPP companies who have submitted detailed applications for transient employment with OSHA to participate on Onsite teams. They receive special SGE training from OSHA and are sworn in as government employees who are covered by the government code of ethics. Although the work site cannot prohibit any regular OSHA employee from being on a Team, they can disallow any SGE from participating. Reasons for such a refusal include previous relationships and potential conflicts of interest due to business issues and proprietary issues. SGEs are also precluded from performing evaluations at any workplace that they have mentored or those in which they have a significant financial interest through direct stock ownership.

In addition to scheduling the Team, the OSHA VPP Manager or Coordinator will make logistical arrangements for travel and lodging. They may ask for some recommendations for the lodging and restaurants in the area of the workplace. The VPP Manager or Coordinator will also request a medical access order from the OSHA national office. The medical access order will allow OSHA to review employee medical records to verify the information on the OSHA logs and other supporting materials.

OSHA may also request from the applicant electronic copies of all hazard control programs. These would include, as examples, control of hazardous energy (lockout/tagout), chemical hazard communication (HazCom), confined-space entry, hot work, and powered industrial trucks (fork trucks). These would be shared with the rest of the Team before the Onsite so that they can become familiar with the programs before they start the evaluation. This is not yet a standard practice and it is not a requirement of the VPP. Some VPP Managers do not follow this practice because they realize that many of the programs may change from the time they are provided to OSHA and the time of the Onsite.

Now let us address what the applicant has to do, and should not do, between the time of the setting of the schedule and the actual Onsite. First and foremost, all at the workplace must maintain the focus on safety and health and not become over anxious about the Onsite. Although there is usually a great desire to clean up the facility so that it looks like it did when it was new, that is not necessary since OSHA may not be impressed and may actually become skeptical about what you were trying to cover up. However, that does not mean that the applicant should not continue to correct known hazards and to maintain an orderly and clean workplace. Many in the safety and health field make their initial judgment about the safety and health of a workplace from their first observations. If they see a disorderly workplace or heavy dust and dirt with aisles and doors blocked, their impression will not be positive and the applicant will have to work up from there. If they observe an orderly clean workplace, their impression will be more positive and that will set the tone for the rest of the evaluation.

THE OPENING CONFERENCE

The evaluation will start with an opening conference. The purpose of the opening conference is to provide an opportunity for the OSHA Team Leader to explain the evaluation process. First, we will discuss who should attend the opening conference and then we will discuss the typical agenda for the opening conference. At a minimum, the plant manager with the rest of the management staff should be represented, representing each of the operational units and other departments such as maintenance, human resources, and quality control. The hourly employees must also be represented. The hourly employees may be represented by the safety committee members. If there is a specific VPP committee or steering team, they should also be present. For represented sites, members of the union executive committee or shop steward should also be present. In addition, as many other employees should be invited as production will allow.

The opening conference should begin with a welcome by the plant manager and introduction of the principal attendees and the OSHA Team. The next topic should be a brief discussion of emergency response procedures including how to recognize a call to evacuate and identifying the evacuation staging areas. The floor should then be turned over to the OSHA Team Leader who will explain the VPP evaluation process. A detailed explanation will follow after this discussion of the agenda. Once the OSHA Team Leader has discussed the evaluation process, the OSHA Team should be given a detailed introduction of the workplace. This introduction should include the organizational structure, the breakdown of the workforce including contractors, a description of the shift work, a layout of the physical plant, the type of work, products produced, and significant hazards that the team should be aware of such as hazardous chemicals or noise and required personal protective equipment. Since this information will be familiar to the company employees, most of them can be excused at that time to return to work. The OSHA team should then be required to undergo whatever training is required of contractors. If the contractor training is very extensive, they can be allowed to get an abbreviated version since they will not be without a company escort at any time.

EVALUATION FORMAT

The evaluation will consist of three methods of data collection. Let us consider the evaluation tools as a three-legged stool. Without each of the three legs, the stool will fall. The three legs of the evaluation are the visual observations made during the audit tour, the review of all of the supporting safety and health documentation, and the interviews of the employees and managers.

We will discuss each of the legs and what the applicant must do to prepare for the Onsite. Some believe, including the authors, that the interviews are the most important factor to consider when performing an Onsite. That is in line with the authors' belief that of each of the elements of an effective safety and health management system, management leadership and employee involvement are the most important and critical to a successful system. Let us consider an analogy of a car's longevity. We are all interested in keeping our cars running as long as possible. If we buy the most expensive car with the best repair record and fail to maintain it, it will fail prematurely. Conversely, if we buy the least expensive car with a poor repair record and maintain it religiously, it will last longer than expected. That analogy is relevant to the culture of safety and health as demonstrated by management's leadership and the sense of ownership of the employees. A strong safety and health culture will generally prevail over poor conditions. That is because the positive culture will tend to properly address the observed weaknesses. Again, conversely, the best control programs and cleanest work environment will deteriorate if the culture is not strong enough to maintain them.

Since the culture for safety and health in the workplace is the foundation for the entire system, OSHA places a great deal of emphasis on the interview process of the Onsite as a means of determining the effectiveness of that culture. Therefore, it is imperative that all employees are prepared for the interviews. First, let us emphasize that when we discuss employees, we are including company hourly and salaried employees, as well as contractors working at the applicant's workplace. OSHA will select to interview representatives from each group, including the senior applicant manager and the senior union official, if applicable. All interviews are confidential and one-on-one between OSHA and the employee.

There are two distinct types of interviews that OSHA performs during the Onsite. One is the informal and the other is the formal interview. First we will discuss the informal interview. That one requires less preparation for the interviewee. The purpose of the interview is to help OSHA determine the knowledge of the employee of the hazards of the work being performed and the training they receive to enable them to protect themselves. Members of the OSHA Team will occasionally stop during their audit of the workplace to speak to randomly selected employees. They will usually stop to speak to an employee that is performing an interesting or extremely hazardous task, or is using personal protective equipment such as a respirator. They will begin by introducing themselves and ask the employee if they may ask some questions. They will usually ask the employee how long he or she has worked at the facility and to describe what he or she is doing. The Team will be interested in any task-specific training and training specifically related to the personal protective equipment used. For example, if a respirator is used, or even nearby, the employee will be asked to explain how it is inspected, donned, cleaned, maintained, and stored. If the employee is working with chemicals, the Team may ask about the MSDS and chemical-related training received by the employee. The Team may also ask informally about what the employee is supposed to do in case of an emergency, and describe his or her feelings about management's commitment to employee safety and health. Since this is a VPP evaluation, the Team will also ask about the VPP.

The Team will also select a number of employees to be interviewed formally. There is no rule as to how many employees will be formally interviewed. The number of formal interviews depends on the total number of employees at the workplace and the size of the Team. A rule of thumb is that each Team member will interview about 8–12 employees. If the workplace is small and there are only a few employees, the Team may elect to interview all employees. If the workplace is very large, the Team members will limit their interviews to the time available. The main consideration is to interview a representative sample of all employees to include all production and support areas and all tasks. Support areas include receiving, warehousing, shipping, maintenance, and office administration.

For the formal interviews, the Team will usually request an organization chart identifying all managers and supervisors, and a list of all employees working during the Team's presence at the workplace. It should be detailed to identify the type of work performed by each employee and the unit of production. If there are fixed shifts, provide the Team with a list of all employees on each shift. The Team will generally limit interviews for rotating shifts to those employees on the day shift. The rational for that is that each employee on the day shift could represent all of the shifts.

The Team Leader will go through the organization chart and employee listing and use an informal random selection process to identify the employees to be interviewed. OSHA has recognized that many people feel very uncomfortable in an interview setting and they recognize that some selected employees may decline the offer. That is not an issue, unless the declination is pervasive. In fact, the authors have experienced such a situation. During a Merit reevaluation, all selected union employees refused the offer to be interviewed. In addition, the union leader, when asked to speak to the Team Leader stated that he "did not want to be involved." Needless to say, the company did not achieve VPP Star and in fact was asked by OSHA to withdraw from the VPP.

The interviews will be scheduled to limit disruptions to production. The applicant will have to provide enough rooms to the Team so that Team members can do their interviews concurrently but separately. The rooms only have to be large enough for two people to sit comfortably for about 15–20 minutes. The Team will expect the applicant to assign someone to schedule the interviews and to arrange for the interviewees to arrive on time at the interview location.

Now that we have discussed the logistics of the formal interviews, let us discuss the interview itself. First of all, the interview is not a test, and all employees can and should be prepared in advance. The VPP Policies and Procedures Manual, available on the OSHA website, contains the recommended interview questions that the Team may ask. They are contained in OSHA Directive CSP 03-01-003, Appendix F, at the following website: http://www.osha.gov/OshDoc/Directive_pdf/CSP_03-01-003.pdf.

These questions can be shared with the employees to review, but it must be emphasized that these are perception-type questions and not pass–fail questions. It must also be emphasized the interviewers are trained and able to work with the interviewees to ensure that they understand the question. The most significant areas of the questions include the interviewees' feelings about their involvement and sense of ownership of their safety and health, and management's commitment and sincerity about their safety and health, the training they receive, and their knowledge of what to do in case of emergency.

One group of questions that applicants prepare their employees to address are those directly related to the VPP. Do employees understand the VPP and what it means to them? Many believe that if employees are not aware of what the VPP is and cannot describe in detail what it means to them and the company, it will have a negative effect on the outcome of the evaluation. In fact, that is the furthest from the truth. The VPP Policies and Procedures Manual and the VPP *Federal Register* only require applicants to notify employees about the VPP application and what the VPP is. This obligation to notify can easily be demonstrated by records of training and visual cues such as posters and banners. Since the knowledge of the VPP is not as protective as say the lockout program, it is not critical if the employees are not that well versed in the program. In other words, not knowing the details of the VPP is not going to protect the employee from harm, but not knowing the lockout program may result in harm.

Another concern is what if the Team interviews a disgruntled employee. It is amazing how even with random selection, Team Leaders invariably select at least one such employee. To address this, most Team Leaders use an informal ratings scale to eliminate the worst and best interviews. This is similar to the older scoring method used for judging international ice skating contests. The highest and lowest scores of the judges are thrown out and the remaining eight scores are averaged. If many of the interviews are negative, the Team will have to reassess the quality of the safety and health management system.

A second leg of the evaluation is referred to as the workplace walkthrough. Since the VPP is a cooperative program, OSHA does not refer to this phase as an inspection. In many respects it is similar to an inspection, without the citations, violations, and penalties. The Team will tour the entire workplace to observe physical conditions, operations, and employee behaviors. Unlike a typical OSHA inspection, unless the workplace is small, the Team will not usually tour it in its entirety. The Team will make it a point to observe each different operation and area. When there are multiple similar production lines in operation, the Team will select some but not necessarily all of them.

The applicant should always attempt to control the walkthrough, remembering that the Team will ultimately call the shots. Determine a preferred tour route prior to OSHA's visit. This usually follows the production flow, with tangential stops at maintenance and other support areas. During the tour, try to encourage OSHA to follow your route, but expect them to decide where and when they want to go. Remember that a closed or locked door is an invitation to the Team to look behind the door. Just prior to the Onsite, walk the tour yourself and attempt to identify all conditions that the Team may determine are unsafe. These should be corrected or tagged with appropriate hazard warnings.

During the preliminary walkthrough, make a special effort to identify those hazards and/or violations that may be considered low hanging fruit. Contrary to what some believe, the Team will not lighten up once they find their first unsafe condition. Low hanging fruit includes those hazards and violations that are very common but still are found in most workplaces. Examples include unlabeled secondary-use chemical containers (spray bottles), electrical circuits and service disconnects not adequately labeled, missing knockouts on electrical junction boxes and service panels, blocked egress and exit doors, emergency lights not working, emergency equipment such as fire extinguishers and eyewashes in poor condition and not inspected, damaged electrical cords, lack of ground fault circuit interrupters (GFCI) for outlets close to a water source, flammables not stored properly, damaged flammable storage cabinets, or inadequately protected operating equipment. This is just a brief listing of those hazards that are very easy to observe and control. Routine inspections should pick them up, but there should be a special inspection tour prior to the Team's visit.

Try to limit the size of the tour by identifying key individuals that will accompany the Team on their initial walkthrough. They should include representatives from safety, union or hourly employees, management, and maintenance. The maintenance representative should be prepared to immediately correct those findings that are easy to address. These include clearing blocked aisles, cutting damaged electrical cords, and replacing missing electrical knockouts. Those items that cannot be immediately corrected should be identified with a tag that describes the hazard, and either takes the offending item out of service or provides a warning.

During the Onsite evaluation, the Team will point out to their escorts what their concerns are and may discuss recommended corrective actions. The Team will also point out those things that impress them positively. The company tour guides should photograph what the Team points out and have someone document all findings, trying to keep a close relationship between the photos and the notes. Each day, at the end of the tour, the notes should be expanded to a tracking system such as an Excel table that includes at least the following fields: date, location, finding, recommended corrective action, person responsible for correction, assigned correction date, final corrective action, and final correction date. In addition and, if possible, identify the hazard with a warning tag. All findings must be accounted for, even if the correction was immediate as in the cutting of a damaged electrical cord. Pictures should also be taken of all corrective actions. They will represent the verification of correction that the Team requires.

The third leg of the evaluation is the document review. This will usually occur on the second and following days. The document review will also include additional tours of the work site by the program reviewers. The documents to be reviewed include but are not limited to all of those hazard control programs in operation at the work site, the OSHA logs for the current year and the three full calendar years prior to the visit, inspection records, first-aid logs, medical records including surveillance, training records and programs, committee minutes, performance evaluations, disciplinary actions, job hazard analyses, work permits, work orders, and any others that support the safety and health management system.

The Team Leader will assign the various team members to review selected records and programs. Generally, the Team Leader reviews the personnel documents such as performance evaluations and disciplinary actions. The safety professional generally reviews programs such as control of hazardous energy and electrical safety related work practices. The industrial hygienist generally reviews the programs associated with chemicals and occupational health such as chemical hazard communication and industrial hygiene sampling. There is no absolute to this breakdown, and there will usually be some crossover. Other team specialists may review documents for such programs as process safety management and cranes.

We will offer some examples of how these reviews occur. The first example is the program to control the unexpected release of hazardous energy. Most workplaces are required to have an effective lockout/tagout program (LOTO) for this purpose. Some may go by different names such as zero energy state or LOTOTO (lockout/tagout/tryout), but they all follow the same principles. Those principles include identifying the hazards throughout the workplace, developing methods to control the energy, providing the equipment needed, training all authorized and affected employees, and making the program evergreen by regular reviews and audits. The team member assigned to review the LOTO program will usually start by reading the program itself and relating it to the observations made during the initial walkaround of the workplace. They may also review the associated employee training program. They will then identify at least one piece of equipment that has a LOTO procedure and ask that an authorized employee simulate an actual lockout of that equipment. The employee should then describe and show the Team member how that equipment is locked out. It should start with a review of the equipment specific procedure, notification to all affected employees, inspection of the equipment, turning off the equipment, identifying each type of energy to be controlled and where the control point is, and pointing out the specific lockout devices to be used. The employee should then walk down the equipment and explain where and how each control device would be placed. The employee should then describe the testing of each energy source to ensure that it actually is controlled. Once the demonstration is complete, the employee should reverse the procedure and end with the simulated reactivation of the equipment.

Another program that is usually found in most workplaces is to control hazards associated with powered industrial trucks (PIT) or fork trucks. This review would start with a review of the PIT program and the daily inspection records. The reviewer will also interview several operators who should be able to describe their training, the rules of operation, and the requirements for regular inspections. This review will also include a discussion of vehicle maintenance, especially the procedure for charging batteries and changing propane tanks. The PIT training program will also be reviewed, as well as the test results. The reviewer will also be interested in retraining requirements and any associated disciplinary programs for infractions such as speeding or driving a PIT without wearing a seat belt.

At the end of each day, the Team will usually caucus to discuss all of their findings, and they will then meet with the company representatives for a daily out-brief to recap the day's activities. The Team Leader will review all observations and provide the company with an opportunity to ask any questions they may have about the findings or the process. The Team will also inform the company what their plans and schedules are for the following day. Although the company representatives are encouraged to ask questions during all phases of the evaluation, this is an opportunity to review all questions and concerns in detail. Misinterpretations or misunderstandings of the Team's comments must be clarified, and those hazards that have already been controlled must be pointed out. It is also critical that all of the Team's findings are documented by the company so that they can all be addressed in the shortest amount of time. The Team will also discuss their plans for the following day. The next day usually starts with a brief meeting. That is an opportunity for the company representatives to ask any questions about the previous day's activities and findings. It also provides the company an opportunity to demonstrate their commitment to the VPP by providing evidence that some of those findings have already been addressed. This could be done by showing pictures of the corrections and reviewing the tracking sheet of the findings.

At the conclusion of the evaluation, the Team will usually reconvene in a private meeting room to discuss all of their findings, write their report, and reach a consensus for their recommendation. They will most likely use their own computer to write the report, but they may ask for the use of an LCD computer projector and a full keyboard to connect to their computer to facilitate the writing process.

The recommendation process is usually straightforward, with the Team Leader simply asking each Team member what their own recommendation for participation is. We will discuss the possible recommendations shortly. There may be some situations where the recommendation is not clearly apparent. Some Team members may be considering Star and some may be considering Merit. To conclude the evaluation, the Team must reach a consensus that all Team members can agree with. These situations may be addressed by the Team Leader in different ways. Generally, the Team Leader will reserve his or her recommendation until the end to avoid swaying the rest of the Team based on the leader's more extensive experience.

One solution that has been used is to simply have each Team member support his or her recommendation and the factors that were considered. Once all positions have been presented, the other team members may change their own positions. The Team Leader would also present his or her position with the rational for it. Another method that has been used when the recommendation is not clear in the past has been to ask each Team member to submit a secret recommendation. Once all are received, the Team Leader will read them and hopefully the Team does reach agreement. An important consideration is that if the Team does have difficulty reaching agreement, they will most likely recommend Merit instead of Star. There is another option that is available but rarely used.

The Team members may defer their recommendation for up to 90 days to allow the work site to address the more significant concerns. This is usually used when the Team feels confident that the concerns can be addressed within 90 days. This option is used primarily to allow for improvement in one or more of the elements of the safety and health management system. It is not for the correction of hazards. One example may be to have the work site improve its inspection program by providing site-specific hazard recognition training to all those performing inspections. The expectation is that the training will be completed within 90 days and then the work site would be recommended for Merit to demonstrate the improvements in the inspections over a complete year. This also helps to demonstrate management's commitment to the VPP.

The final option is even more rarely used. That is when the Team is so discouraged in their findings that they must ask the applicant to withdraw the application. To visualize when this recommendation would be appropriate, imagine the most unsafe or unhealthful working conditions you can. Another reason would be because the employees demonstrate a total lack of involvement in the safety and health activities or through the interviews describe a workplace that encourages employees to take safety shortcuts and where required personal protective equipment use is nonexistent. Clearly, these are workplaces that do not deserve to be grouped with industry leaders in safety and health. These work sites are requested to withdraw their applications. Should they refuse, the OSHA assistant secretary will be asked by the OSHA regional administrator to order a return of the application.

Once the Team has reached a consensus on the recommendation, they will complete the VPP Onsite evaluation report. The report will consist of a narrative description of the work site, the number of employees, and unions represented, the type of work performed, and the products produced. It will also compare the work site's 3-year injury and illness rates to the BLS rates for the appropriate NAICS code. This is a critical comparison because it is the only objective measure considered in the VPP evaluation process. To achieve Star recognition, the site's TCIR and DART 3-year rates must be below the industry rates for any of the most recent last 3 years reported by BLS for the closest NAICS code. Both the TCIR and DART rates must be below for the same year. If either one or both site rates are above the BLS data, the site may be recommended for Merit with one or more Merit goals that when completed will enable the site to realize rate reductions to meet the Star requirements. Merit goals related to rate reductions cannot exceed 2 years. If the rates cannot be reduced to below the BLS rates within that 2-year period, the work site will be requested to withdraw from the VPP. It must be programmatically and mathematically feasible to reduce the site's 3-year rates within 2 years.

Once the draft report is completed, the OSHA Team will be ready to conclude the evaluation with the closing conference. Like the opening conference, the work site should have as many associates as possible present to hear the results of the evaluation. The closing conference will follow a very brief agenda of the site's VPP coordinator introducing the Team leader. The Team Leader will then briefly thank all work-site associates for their assistance and hospitality and immediately state the recommendation; a quick recommendation limits any anxiety and suspense. The team's recommendation is then referred to the Regional Administrator for concurrence. The Team Leader will then discuss the Team's findings of each of the major elements of the safety and health management system. The Team Leader will then discuss the next steps in the process. The closing conference should conclude with a brief statement from the senior work-site manager to thank the Team members for their participation in the evaluation process and their assistance in improving the safety and health management system. The senior manager should also thank those associates that worked to prepare the application and the work site for the VPP and the evaluation. A statement of commitment to the VPP and the implementation of all identified opportunities for improvement is also appropriate. The evaluation is then over with the exception of the final farewells and handshakes.

Exhibit 8.1 presents a sample of the OSHA onsite evaluation report template.

Exhibit 8.1 Sample OSHA Onsite Evaluation Template

VPP SITE REPORT

Recommending

STAR APPROVAL

For

Your Company

Anytown, USA

Evaluation Dates

July 7–11, 2008

Report Date

July 11, 2008

Evaluation Team

OSHA Team Leader

OSHA Industrial Hygienist

SGE, Safety Specialist

SGE, Safety Specialist

I. Purpose and Scope of Review

An onsite VPP approval review was conducted at Your Company from July 7th through July 11th, 2008. The OSHA VPP Review Team (Team) consisted of:

OSHA Team Leader

OSHA Industrial Hygienist

SGE, Safety Specialist

SGE, Safety Specialist

II. Methods of Data Collection

The information for this report was obtained from the site's VPP application; documentation reviewed on-site, interviews with employees, and a walk-through of the plant. All areas of the plant were covered in at least one walk-through.

III. Employees at the Worksite

There are 476 employees currently working on-site. In addition, 27 contractor employees are located on-site performing security, cafeteria, janitorial and warehouse functions. There is no collective bargaining agent representing the employees on-site. Formal interviews were conducted with 24 employees and no contractors. In addition, informal interviews were conducted with 67 employees and 9 contractors.

IV. The Worksite

The site is properly classified under the Standard Industrial Classification (NAICS) code XXXXXX for the manufacturing of widgets. The site was built in 1974. It consists of three production lines, a quality control lab for structural analyses, a warehouse, and administrative offices. The one building is located in an industrial park. Primary process is manufacturing of widgets. The site is not covered under OSHA's Process Safety Management (PSM) standard for any of the chemicals used or stored in the process or plant.

V. Worksite Hazards

The safety hazards located on this site include: mechanical equipment, confined spaces, hazardous energy sources, operation of moving equipment such as powered industrial trucks. Occupational health hazards include: elevated noise levels; and material handling.

VI. Injury and Illness Rates

For the period 2005–2007, the site's:

- Total Case Incidence Rate (TCIR) is 3.0 (30% below_the 2006 BLS industry averages for NAICS XXXXXX.
- The Days Away from Work, Restricted Activity or Job Transfer (DART) case incidence rate is 1.6 (48% below_the 2006 BLS industry averages for NAICS XXXXXX).

VII. OSHA Activity

There has been no OSHA inspection activity at the site in any of the last five years. The site has a positive relationship with the local OSHA office. The site contacted the OSHA Regional Office regarding participation in the VPP program.

VIII. Elements of the VPP Review/Program Changes

The OSHA VPP Review team examined each of the required elements of the site's safety and health management system and found them to be consistent with the high quality of VPP programs. The site meets all VPP requirements and all OSHA standards are appropriately covered. For specifics on the individual site program elements, consult the VPP Site Worksheet. In addition, the site has submitted the following summary of the SHE improvements initiated since the last evaluation:

A review of the OSHA 300 logs was made. The following are the total incidence and lost workday case rates since 2005:

Year	Hours	Total # of Cases	TCIR	Activity or Job Transfer	DART Rate
2005	952000	15	3.2	9	1.9
2006	960000	16	3.3	9	1.9
2007	930000	11	2.4	5	1.1
Total	2842000	42		23	
Three-Year Rate (2005–2007)			3.0		1.6
BLS National Average for 2006 (NAICS XXXXXX)			4.3		3.1
2008 YTD	410000	5	1.1	2	0.5

For the period 2005–2007, the site's:

- Total Case Incidence Rate (TCIR) is 3.0 (30% below_the 2006 BLS industry averages for NAICS XXXXXX.

- The Days Away from Work, Restricted Activity or Job Transfer (DART) case incidence rate is 1.6 (48% below_the 2006 BLS industry averages for NAICS XXXXXX).

The information on the OSHA 300 logs supports the information provided in the application, and the company's first report of injury forms support the data in the logs. The plant Occupational Health Nurse is responsible for the entries to the OSHA 300 log and verified the accuracy of the records. The plant Occupational Health Nurse understands the recordkeeping requirements. Based upon interviews conducted with management and employees, the logs accurately reflect the injury and illness experience at this plant.

There was one temporary employee at the worksite at the time of the team's visit. Injuries or illnesses experienced by temporary employees under the direct supervision of Your Company are recorded on the worksite's OSHA 300 log.

VPP SITE WORKSHEET

Recommending

STAR APPROVAL

For

Your Company

Anytown, USA

Evaluation Dates

July 7–11, 2008

Report Date

July 11, 2008

Evaluation Team

OSHA Team Leader

OSHA Industrial Hygienist

SGE, Safety Specialist

SGE, Safety Specialist

Section I: Management Leadership & Employee Involvement	**Yes or No**	**How Assessed** Interview	Observation	Doc Review
A. Written Safety & Health Management System				
A1. *Are all the elements (such as Management Leadership and Employee Involvement, Worksite Analysis, Hazard Prevention and Control, and Safety and Health Training) and sub-elements of a basic safety and health management system part of a signed, written document? (For Federal Agencies, include 29 CFR 1960.) If not, please explain.* • SHE Documents and Checklist. • Safety Policies and Procedures.	Y			X
A2. *Have all VPP elements and sub-elements been in place at least 1 year? If not, please identify those elements that have not been in place for at least 1 year.* • Can be verified by reviewing the SHE links.	Y		X	X
A3. *Is the written safety and health management system at least minimally effective to address the scope and complexity of the hazards at the site? (Smaller, less complex sites require a less complex system.) If not, please explain.* MR⊘. • SHE Documents and Checklist. • Safety Policies and Procedures.	Y		X	X
A4. *Have any VPP documentation requirements been waived (as per FRN page 656, paragraph F5a4)? If so, please explain.*	N	X	X	X

Section I: Management Leadership & Employee Involvement	Yes or No	How Assessed: Interview	Observation	Doc Review
B. Management Commitment & Leadership				
B1. *Does management overall demonstrate at least minimally effective, visible leadership with respect to the safety and health program (considering FRN items F5 A-H)? Provide examples.* **MR⊘**.	Y	X	X	
• Plant manager & SHE Manager conduct field audits twice a week. • Leadership Team maintains records to track PPE, Housekeeping inspections and compliance metrics by area. • Site conducts multiple surveys to solicit input from employees: Opinion Survey, Employee Roundtables, all employee meetings, and annual SHE program evaluations.				
B2. *How has the site communicated established policies and results-oriented goals and objectives for worker safety to employees?*		X		X
• SHE Website on intranet, routine all employee meetings, new employee orientation training, weekly/monthly safety meetings				

Section I: Management Leadership & Employee Involvement	Yes or No	How Assessed: Interview	Observation	Doc Review
B3. *Do employees understand the goals and objectives for the safety and health program? if not, please explain.*	Y	X		
B4. *Are the safety and health program goals and objectives meaningful and attainable? Provide examples supporting the meaningfulness and attainability (or lack-there-of if answer is no) of the goal(s). (Attainability can either be unrealistic/realistic goals or poor/good implementation to achieve them.) (See: TED Chapter 3 II C1a)* • Site SHE goals and objectives. • Site reviews progress on SHE goals and objectives on a monthly basis. • Site has identified historical hazards specific to each month of the year. Improvement plans are based on historical trends as well as BST data.	Y			X
B5. *How does the site measure its progress towards the safety and health program goals and objectives? Provide examples.* • Site management reviews goals and objectives quarterly and compliance metrics weekly.		X		X

		Yes or No	How Assessed: Interview	Observation	Doc Review
Section I: Management Leadership & Employee Involvement					
C. Planning					
C1.	*How does the site integrate planning for safety and health with its overall management planning process (for example, budget development, resource allocation, or training)?*		X		X
•	The SHE manager is responsible for SHE funding and site project approval for the 3 yr investment plan. The site has an established 3 yr plan for routine SHE expenses. This is reviewed on a bi-weekly basis with the site leadership team. The SHE Manager is responsible for managing this funding.				
C2.	*Is safety and health effectively integrated into the site's overall management planning process? If not, please explain.*	Y	X		X

	Yes or No	How Assessed: Interview	Observation	Doc Review
Section I: Management Leadership & Employee Involvement				
D. Authority and Line Accountability				
D1. *Does top management accept ultimate responsibility for safety and health in the organization? (Top management acknowledges ultimate responsibility even if some safety and health functions are delegated to others.) If not, please explain.* **MR⊘**. • Site leadership holds weekly SHE progress reviews. • The Leadership team along with direct reports review goals and objectives on a quarterly basis: all employee goals include SHE.	Y	X		
D2. *How is the assignment of authority and responsibility documented and communicated (for example, organization charts, job descriptions)?* • Organization chart.				X
D3. *Do the individuals assigned responsibility for safety and health have the authority to ensure that hazards are corrected or necessary changes to the safety and health management system are made? if not, please explain.* **MR⊘**. • Through safety committee overview, annual self-assessments, tracking system, quarterly Round Tables, quarterly All Employee Meetings, SHE Manager support.	Y	X		X

Section I: Management Leadership & Employee Involvement	Yes or No	How Assessed: Interview	Observation	Doc Review
D4. *How are managers, supervisors, and employees held accountable for meeting their responsibilities for workplace safety and health? (Annual performance evaluations for managers and supervisors are required.)* • Quarterly annual performance reviews.	▨			X
D5. *Are adequate resources (equipment, udget, or experts) dedicated to ensuring workplace safety and health?* **MR⊘**. • Annual budget for the past three years was about $1.5 million.	Y			X
D6. *Is access to experts (for example, Certified Industrial Hygienists, Certified Safety Professionals, Occupational Nurses, or Engineers), reasonably available to the site, based upon the nature, conditions, complexity, and hazards of the site? If so, under what arrangements and how often are they used?* • Certified Industrial Hygienist, Certified Safety Professional, 1 OHST, 1 Certified Occupational Health Nurse, Mechanical & Industrial Engineers.	X			X

Section I: Management Leadership & Employee Involvement	Yes or No	How Assessed: Interview	Observation	Doc Review
E. Contract Workers				
E1. *Does the site utilize contractors? Please explain.*	Y		X	X
• Site utilizes contractors for capital project execution.				
E2. *Were there contractor's onsite at the time of the evaluation?*	Y		X	
E3. *When selecting onsite contractors, how does the site evaluate the contractor's safety and health programs and performance (including rates)? (See: TED Chapter 3 IV 3-19)*				X
• Site contractor program requires SHE evaluation of the contractor's safety record and program annually.				
E4. *Are contractors and subcontractors at the site to maintain effective safety and health programs and to comply with all applicable OSHA and company safety and health rules and regulations? If not, please explain.*	Y			X
• Periodic SHE evaluation.				
E5. *Does the site's contractor program cover the prompt correction and control of hazards in the event that the contractor fails to correct or control such hazards? Provide examples.* **MR⊘**.	Y			X
• The site has removed contractors for unsafe activities.				
• Interviewed contractors stated that if they were to work unsafely or violate a safety rule they would be disciplined up to removal from the site.				
E6. *How does the site document and communicate oversight, coordination, and enforcement of safety and health expectations to contractors?*				X
• Contractor Policy, Field observations, Monthly meetings, Contractor orientation.				

Section I: Management Leadership & Employee Involvement	Yes or No	How Assessed: Interview	Observation	Doc Review
E7. *Have the contract provisions specifying penalties for safety and health issues been enforced, when appropriate? If not, please explain.* • The site has documentation of removing contractors that had declining safety performance trends.	Y			X
E8. *How does the site monitor the quality of the safety and health protection of its contract employees?* • Contractor Policy, Field observation, Monthly meetings, Contractor orientation.				X
E9. *If the contractors' injury and illness rates are above the average for their industries, does the site have procedures that ensure all employees are provided effective protection on the worksite? If not, please explain.* • Contractors are given an opportunity to improve or be removed from site. • Contractor policy specifies TCIR and DART rates.	Y			X
E10. *Do contract provisions for contractors require the periodic review and analysis of injury and illness data? Provide examples.* • Injury and Illness data trends are evaluated on a case by case basis with the contract firm.	Y			X
E11. *Based on your answers to the above items, is the contract oversight minimally effective for the nature of the site? (Inadequate oversight is indicated by significant hazards created by the contractor, employees exposed to hazards, or a lack of host audits.) If not, please explain.* **MR⊘**.	Y		X	X

Section I: Management Leadership & Employee Involvement	**Yes or No**	**How Assessed** Interview	Observation	Doc Review
F. Employee Involvement				
F1. *How were employees selected to be interviewed by the VPP team?*				
• Site provided employee roster by department for random selection by the VPP Team Leader .				
F2. *How many employees were interviewed formally? How many were interviewed informally?*				
• 24 formal and 67 informal.				
F3. *Do employees support the site's participation in the VPP Process?* **MR⦸**.	Y	X		
F4. *Do employees feel free to participate in the safety and health management system without fear of discrimination or reprisal? If so, please explain.* **MR⦸**.	Y	X		
F5. *Please describe at least three ways in which employees are meaningfully involved in the problem identification and resolution, or evaluation of the safety and health program (beyond hazard reporting). (See: FRN Chapter 3 Paragraph II.C.1.b)*		X		X
• Wellness team, ERT/EMT, BST, Safety Rules Committee, JHAs.				

		Yes or No	How Assessed: Interview	Observation	Doc Review
Section I: Management Leadership & Employee Involvement					
F6.	*Are employees knowledgeable about the site's safety and health management system? If not, please explain.*	Y	X		
F7.	*Are employees knowledgeable about the VPP program? If not, please explain.*	Y	X		
•	The employees strongly support the VPP.				
F8.	*Are the employees knowledgeable about OSHA rights and responsibilities? If not, please explain.*	Y	X		
F9.	*Do employees have access to results of self-inspection, accident investigation, appropriate medical records, and personal sampling data upon request? If not, please explain.*	Y	X		
•	Medical Department, Safety Alerts and Communications, Intranet.				

Section I: Management Leadership & Employee Involvement

Merit Goals (*Include cross reference to section, subsection, and question, e.g., I.B2*)

1.

2.

90-Day Items (*Delete this section for final transmittal to National Office*)

1.

2.

Best Practices

1.

2.

Comments including Recommendations (*Optional*)

1.

2.

		Yes or No	How Assessed: Interview	Observation	Doc Review
Section II: Worksite Analysis					
A. Baseline Hazard Analysis					
A1.	*Has the site been at least minimally effective at identifying and documenting the common safety and health hazards associated with the site (such as those found in OSHA regulations, building standards, etc., and for which existing controls are well known)? If not, please explain.* **MR⊘**.	Y	X		X
•	Historic injury trends, BST data, IH program, JHAs.				
A2.	*What methods are used in the baseline hazard analysis to identify health hazards? (Please include examples of instances when initial screening and full-shift sampling were used. See FRN page 45657, F5.B.2.b)*		X		X
•	Medical evaluations, Baseline IH & JHAs, MOC.				
A3.	*Does the site have a documented sampling strategy used to identify health hazards and assess employees' exposure (including duration, route, and frequency of exposure), and the number of exposed employees? If not, please explain.*	Y			X
•	CIH develops the strategy and an annual sampling plan.				
A4.	*Do sampling, testing, and analysis follow nationally recognized procedures? If not, please explain.*	Y			X
•	Site uses NOISH or OSHA procedures and certified laboratory to analysis samples.				

Section II: Worksite Analysis		Yes or No	How Assessed: Interview	Observation	Doc Review
A5.	*Does the site compare sampling results to the minimum exposure limits or are more restrictive exposure limits (PELs, TLVs, etc.) used? Please explain.*	Y	X		X
•	Site follows the lowest applicable regulatory or recommended occupational exposure limit (OEL).				
•	They also use ½ the OSHA PEL as a trigger level for action.				
A6.	*Does the baseline hazard analysis adequately identify hazards (including health) that need further analysis? If not, please explain.*	Y			X
•	Baseline hazards are reviewed prior to start up using the Management Of Change process, requiring a sign-off before equipment or material usage.				
A7.	*Do industrial hygiene sampling data, such as initial screening or full shift sampling data, indicate that records are being kept in logical order and include all sampling information (for example, sampling time, date, employee, job title, concentrated measures, and calculations)? If not, please explain the deficiencies and how they are being addressed.*	Y			X
•	Computerized system and hard copies are logged and kept in accessible locations. Information includes all of the information from the date of sampling, calibrations and methods.				

Section II: Worksite Analysis		Yes or No	How Assessed: Interview	Observation	Doc Review
B. Hazard Analysis of Significant Changes					
B1.	*When purchasing new materials or equipment, or implementing new processes, what types of analyses are performed to determine their impact on safety and health? Is it adequate?*	Y			X
•	MSDS, MOC, PSSR reviews.				
B2.	*When implementing/introducing non-routine tasks, materials or equipment, or modifying processes, what types of analyses are performed to determine their impact on safety and health? Is it adequate?*	Y	X		X
•	MOC, JHA, IH Baseline Monitoring.				

		Yes or No	How Assessed: Interview	How Assessed: Observation	How Assessed: Doc Review
Section II: Worksite Analysis					
C. Hazard Analysis of Routine Activities					
C1.	*Is there at least a minimally effective hazard analysis system in place for routine operations and activities?* **MR⊗**.	Y			X
•	JHA, MOC.				
C2.	*Does hazard identification and analysis address both safety and health hazards, if appropriate? If not, please explain.*	Y			X
C3.	*What hazard analysis technique(s) are employed for routine operations and activities (e.g., job hazard analysis, HAZ-OPS, fault trees)? Are they adequate?*	Y			X
•	JHA.				
C4.	*Are the results of the hazard analysis of routine activities adequately documented? If not, please explain.*	Y			X

	Yes or No	How Assessed: Interview	Observation	Doc Review
Section II: Worksite Analysis				
D. Routine Inspections				
D1. *Does the site have a minimally effective system for performing safety and health inspections (i.e., a minimally effective system identifies hazards associated with normal operations)? if not, please explain.* **MR⊘**. • Supervisor scorecard. • Weekly site audits by hourly employees.	Y	X	X	X
D2. *Are routine safety and health inspections conducted monthly, with the entire site covered at least quarterly (for construction: entire site weekly)?* • Monthly self audits covering the entire site.	Y			X
D3. *How do inspections use information discovered through the baseline hazards analysis, job hazard analysis, accident/incident analysis, employee concerns, sampling results, etc.?* • Identified opportunities for improvement are identified in previous inspection reports, BST data and historical injury trend analysis along with corporate safety alerts are incorporated into area inspections.				X
D4. *Are those personnel conducting inspections adequately trained in hazard identification? If not, please explain.* • All employees are trained in BST and SHE SOPs.	Y	X		X

Section II: Worksite Analysis		Yes or No	How Assessed: Interview	Observation	Doc Review
D5.	*Is the routine inspection system written, including documentation of results? If not, please explain.*	Y			X
D6.	*Do the written routine inspection reports clearly indicate what needs to be corrected, by whom, and by when? If not, please explain.*	Y			X
•	Documented and tracked in the work order system.				
D7.	*Did the VPP team find hazards that should have been found through self-inspection? If so, please explain.*	Y		X	
•	Hazards found by the team were promptly corrected.				

Section II: Worksite Analysis		Yes or No	How Assessed: Interview	Observation	Doc Review
E. Hazard Reporting					
E1.	*Does the site have a reliable system for employees to notify appropriate management personnel in writing about safety and health concerns? Please describe.*	Y	X		X
•	E-mail, Near Miss reports, suggestion reports, or directly to supervisor.				
E2.	*Do the employees agree that they have an effective system for reporting safety and health concerns? If not, please explain.*	Y	X		
•	Almost unanimous.				
E3.	*Is there a minimally effective means for employees to report hazards and have them addressed? If not, please explain.* **MR⊘**.	Y	X		X
•	E-mail, Near Miss reports, suggestion reports, or directly to supervisor.				

Section II: Worksite Analysis		Yes or No	How Assessed: Interview	Observation	Doc Review
F. Hazard Tracking					
F1.	*Does the hazard tracking system address hazards found by employees, hazard analysis of routine and non-routine activities, inspections, and accident or incident investigations? If not, please explain.*	Y			X
F2.	*Does the tracking system result in hazards being corrected and provide feedback to employees for hazards they have reported. If not, please explain.* • Weekly reviews. • Reporting employees receive e-mail reply on status.	Y			X
F3.	*Does the tracking system result in timely correction of hazards with interim protection established when needed? Please describe.* • Automated tracking system is used.	Y		X	X
F4.	*Does a minimally effective tracking system exist that results in hazards being controlled? If not, please explain.* **MR⦸**.	Y			X

Section II: Worksite Analysis		Yes or No	How Assessed: Interview	Observation	Doc Review
G. Accident/Incident Investigations					
G1.	*Is there a minimally effective system for conducting accident/incident investigations, including near-misses? If not, please explain.* **MR⊘**.	Y			X
G2.	*Are those conducting the investigations trained in accident/incident investigation techniques? If not, please explain.*	Y			X
G3.	*Describe how investigations discover and document all the contributing factors that led to an accident/incident.*				X
•	Through the root cause analysis process.				
G4.	*Were any hazards discovered during the investigation previously addressed in any prior hazard analyses (e.g., baseline, selfinspection)? Please explain.*	N			X

	Yes or No	How Assessed: Interview	Observation	Doc Review
Section II: Worksite Analysis				
H. Safety and Health Program Evaluation				
H1. *Briefly describe the system in place for conducting an annual evaluation.* • Annual review every January. • Performed by SHE Manager with site leadership and employee involvement. • Process follows VPP template.				X
H2. *Does the annual evaluation cover the aspects of the safety and health program, including the elements described in the* ***Federal Register****? If not, please explain.*	Y			X
H3. *Does the annual evaluation include written recommendations in a narrative format? If not, please explain.*	Y			X
H4. *Is the annual evaluation an effective tool for assessing the success of the site's safety and health system? Please explain.* • Recommendations for improvement result in a corrective action plan that is tracked to completion.	Y			X
H5. *What evidence demonstrates that the site responded adequately to the recommendations made in the annual evaluation?* • Degree of use of the automated tracking system and rolling percentage of closure and verification rates.				X

		Yes or No	How Assessed: Interview	Observation	Doc Review
Section II: Worksite Analysis					
I. Trend Analysis					
I1.	*Does the site have a minimally effective means for identifying and assessing trends?* **MR⊘**.	Y			X
I2.	*Have there been any injury and/or illness trends over the last three years? If so, please explain.*	Y			X
•	Hand injuries including cuts and lacerations and ergonomic injuries.				
I3.	*If there have been injury and/or illness trends, what courses of action have been taken? Are they adequate?*	Y	X	X	X
•	Improved PPE.				
•	MOC resulting in revised operating procedures				
•	Ergonomic program improvements.				
I4.	*Does the site assess trends utilizing data from hazard reports or accident/incident investigations to determine the potential for injuries and illnesses? If not, please explain.*	Y			X
•	Key areas of opportunities that have been identified in BST data and historical injury trend analysis.				

Section II: Worksite Analysis
Merit Goals (*Include cross reference to section, subsection, and question e.g., II.B2*)
1.
2.

90-Day Items (*Delete this section for final transmittal to National Office*)
1.
2.

Best Practices
1.
2.

Comments including Recommendations (*Optional*)
1.
2.

Section III: Hazard Prevention and Control	Yes or No	How Assessed: Interview	Observation	Doc Review
A. Hazard Prevention and Control				
A1. *Does the site select at least minimally effective controls to prevent exposing employees to hazards.* **MR⊘**. • Reference baseline monitoring.	Y			X
A2. *When the site selects hazard controls, does it follow the preferred hierarchy (engineering controls, administrative controls, work practice controls [i.e. lockout/tag out, bloodborne pathogens, and confined space programs], and personal protective equipment) to eliminate or control hazards? Please provide examples, such as how exposure to health hazards were controlled.* • Modification of equipment to reduce ergonomic hazards. Sound dampening equipment Work practice controls.	X			X
A3. *Describe any administrative controls used at the site to limit employee exposure to hazards (for example, job rotation).* • On a case by case basis, an employee might be given light duty or moved to another work task to accommodate a personal issue.				X
A4. *Do the work practice controls and administrative controls adequately address those hazards not covered by engineering or administrative controls? If not, please explain.*	Y			X

Section III: Hazard Prevention and Control		Yes or No	How Assessed: Interview	Observation	Doc Review
A5.	*Are the work practice controls (i.e. lockout/tag out, blood born pathogens, and confined space programs) recommended by hazard analyses implemented at the site? If not, please explain.*	Y			X
•	Programs are comprehensive and exceed OSHA standards when appropriate.				
A6.	*Are follow-up studies (where appropriate) conducted to ensure that hazard controls were adequate? If not, please explain.*	Y			X
•	Exhaust ventilation testing, noise dosimeter testing.				
A7.	*Are hazard controls documented and addressed in appropriate procedures, safety and health rules, inspections, training, etc.? Provide examples.*	Y			X
•	SOPs, safety and health programs and training.				
A8.	*Are there written worker safety procedures including a disciplinary system? Describe the disciplinary system.*	Y			X
•	Progressive based on frequency and severity of violation.				
•	Prior to separation is a one-day suspension with pay as a Decision Making Day.				
A9.	*Has the disciplinary system been enforced equally for both management and employees, when appropriate? If not, please explain.*	Y			X

Section III: Hazard Prevention and Control	Yes or No	How Assessed: Interview	Observation	Doc Review
A10. *Does the site have minimally effective written procedures for emergencies (TED 3-16 3h)?* **MR⊘.**	Y			X
A11. *Are emergency drills held at least annually?* • Site alarms are tested monthly. • Semi-annual emergency drill scenarios • Annual evacuation drills. • Monthly ERT/EMT training. • All non-exempt employees receive annual fire extinguisher training.	Y			X
A12. *Does the site have a written preventative/predictive maintenance system? If not, please explain.* • The PM system is documented • Mechanical integrity inspections are documented	Y			X
A13. *Did the hazard identification and analysis (including manufacturers' recommendations) identify hazards that could result if equipment is not maintained properly? If not, please explain.* • Equipment failures that could lead to a hazard are covered through JHAs.	Y			X
A14. *Does the preventive maintenance system adequately detect hazardous failures before they occur? If not, please explain.* • The PM system includes predictive techniques designed to detect failures before they happen. Thermography, oil analysis, vibration analysis.	Y			X

Section III: Hazard Prevention and Control		Yes or No	How Assessed: Interview	Observation	Doc Review
A15.	*How does the site select Personal Protective Equipment (PPE)?*				X
•	PPE selection is based on information contained in manufacturer specifications, material hazard characteristics, MSDSs, site job analysis and employee input.				
A16.	*Do employees understand the limitations and uses of PPE? If not, please explain.*	Y	X		
•	Employees are trained in the use, care, and limitations of PPE.				
A17.	*Did the team observe employees using, storing, and maintaining PPE properly? If not, please explain.*	Y		X	
A18.	*Is the site covered by the Process Safety Management Standard (29 CFR 1910.119)? If not, skip to section B.*	N			X
•	No				
A19.	*Which chemicals that trigger the Process Safety Management (PSM) standard are present?*				
A20.	*Please describe the PSM elements in place at the site (do not duplicate if included elsewhere in the report, such as under contractors, preventive maintenance, emergency response, or hazard analysis).*				

	Yes or No	How Assessed: Interview	Observation	Doc Review
Section III: Hazard Prevention and Control				
B. Occupational Health Care Program and Recordkeeping				
B1. *Describe the occupational health care program (including availability of physician services, first aid, and CPR/AED) and special programs such as audiograms or other medical tests used.* • The site has a COHN trained in occupational health issues. • He conducts audiograms, pulmonary function tests, annual medical exams for respirator usage, pre-placement exams. • He and the EMT squad respond to work place health emergencies and identify conditions for follow up by the SHE staff.		X		X
B2. *How are licensed occupational health professionals used in the site's hazard identification and analysis, early recognition and treatment of illness and injury, and the system for limiting the severity of harm that might result from workplace illness or injury? Is this use appropriate?* • Serve on Wellness team, included in SHEs bimonthly staff meetings, and provides information to workers on how to reduce exposure, such as ergonomics.	Y	X		X
B3. *Is the occupational health program adequate for the size and location of the site, as well as the nature of hazards found here? If not, please explain.*	Y	X	X	X

Section III: Hazard Prevention and Control
Merit Goals (*Include cross reference to section, subsection, and question, e.g., I.B2*)
1.
2.

90-Day Items (*Delete this section for final transmittal to National Office*)
1.
2.

Best Practices
1.
2.

Comments including Recommendations (*Optional*)
1.
2.

		Yes or No	How Assessed: Interview	Observation	Doc Review
Section IV: Safety and Health Training					
A. Safety and Health Training					
A1.	*What are the safety and health training requirements for managers, supervisors, employees, and contractors?*				X
•	All levels of the organization are included and tracked on completion of SHE training.				
A2.	*Who delivers the training?*		X		X
•	Training is provided through various groups. It is available electronically through use of the Intranet CBT system as well as hands on by trainer employees.				
A3.	*How are the safety and health training needs for employees determined?*				X
•	Training requirements follow standards set by OSHA.				
A4.	*Does the site provide minimally effective training to educate employees regarding the known hazards of the site and their controls? If not, please explain.* **MR⊘**.	Y			X
A5.	*What system is in place to ensure that all employees and contractors have received and understand the appropriate training?*				X
•	Automated tracking system.				

Section IV: Safety and Health Training		Yes or No	How Assessed: Interview	Observation	Doc Review
A6.	*Who is trained in hazard identification and analysis?*	▨			X
•	All site employees trained in BST, SHE Manager and Team, supervisors and operational managers.				
A7.	*Is training in hazard identification and analysis adequate for the conditions and hazards of the site? If not, please explain.*	Y			X
A8.	*Does management have a thorough understanding of the hazards of the site? Provide examples that demonstrate their understanding.*	Y			X
•	Ability to describe the process to the team during the site tour of the evaluation, responsiveness to the team's suggestions, quick response to the team's questions.				

Section IV: Safety and Health Training
Merit Goals (*Include cross reference to section, subsection, and question, e.g., I.B2*)
1. 2.

90-Day Items (*Delete this section for final transmittal to National Office*)
1. 2.

Best Practices
1. 2.

Comments including Recommendations (*Optional*)
1. 2.

9

POSTEVALUATION ACTIVITIES

Once the OSHA VPP team leaves the work site, the work begins to continue the VPP process. The team Leader will give the site VPP Coordinator a copy of the draft report to review and comment on. Although there is no official time frame for the company to review the report, the sooner it is reviewed and returned to OSHA the sooner the application process will be completed. The usual expectation is that the comments on the report will be sent to OSHA within 30 days. The review must consider the information in the executive summary and the worksheet that identifies all of the findings of the team, including 90-day items (hazard correction items), Merit goals, comments, and recommendations.

The report must be carefully reviewed in its entirety. The executive summary of the report includes a description of the workplace. It contains a brief description of the facility, the work process and product produced, the number of employees, significant contractor activity, and all unions representing company employees. The executive summary also includes a comparison of the injury and illness rates for the most recent 3 years with the BLS rates for the identified NAICS codes. All recent OSHA inspection activities will be discussed as well as how the workplace has demonstrated its commitment to the VPP. The executive summary concludes with the evaluation team's recommendation for participation. The review of the executive summary of the report should focus on descriptions of the workplace, the number of employees, and the identification of the unions. Special attention must be given to the description of the work process and the products produced to ensure that they are correct and that there are no references that may be considered as proprietary. Once the work site is

Preparing for OSHA's Voluntary Protection Programs. By Brian T. Bennett and Norman R. Deitch
Copyright © 2010 John Wiley & Sons, Inc.

approved as a VPP participant, the report becomes part of the public record and is made available under the Freedom of Information Act.

The executive summary is followed by the recordkeeping section, which includes a detailed table identifying the injury/illness statistics for each of the most recent 3 years. The table also compares the data to the appropriate BLS rates. The BLS rates used for this comparison are those that are most liberal of the most recent of the last 3 years of published rates for both the TCIR and DART. For example, if the company's rates for both TCIR and DART are lower than the most recent corresponding BLS rates, those rates would be used. If either or both of the company's rates are higher than the corresponding BLS rates, the BLS rates for the previous year would be used for the comparison. If that still does not result in company rates lower than the BLS rates, then the next previous year's BLS rates may be used. The limit is 3 years. If the company's TCIR and DART rates are not lower than the BLS rates for any of the 3 previous years, the result would be a possible participation in the Merit program.

The recordkeeping section also describes who maintains the OSHA logs and how they are maintained and who is responsible for certifying them for accuracy. It will also include the team's findings of the review of the OSHA logs, first-aid logs, near-miss reports, and reviewed employee medical records.

The report should describe the impact of any significant changes (management, mergers, corporate buy-outs, etc.), events (fatality, catastrophe, accident, complaints, etc.), and steps taken to ensure or restore employee safety and health. The description of any corporate changes should focus on how the change has or may affect the work site. Some such changes may result in increased corporate safety and health support, and some may result in reduced corporate support. Regardless of the situation, the report should explain how the work site has or will address those changes. The description of any significant events should be described with caution to avoid disclosing any potentially damaging information. If OSHA was already involved, they will have the details. If they were not already involved, avoid providing information other than a description of what happened.

The recordkeeping section is followed by the VPP onsite evaluation worksheet. The worksheet contains a group of tables and has four sections that address each of the major elements of the safety and health management system (system) (management leadership and employee involvement, work-site evaluation, hazard prevention and control, and safety and health training). Each section of the worksheet contain questions that address each of the subelements of the System. Most of the questions require a "Yes" or "No" answer with a check mark in one or more of three columns to identify if the answer was arrived at by reviewing "interviews," "observations," or "document reviews." Although all questions allow the team to provide their observations in bullet format, several actually require a narrative answer and several require examples. One example of a question that requires a narrative answer is: "*Please describe at least three ways in which employees are meaningfully involved in the problem identification and resolution, or evaluation of the safety and health program (beyond hazard reporting).*" Another is: "*Does the site have a reliable system for employees to notify appropriate management personnel in writing about safety and health concerns? Please describe.*"

All entries in the worksheet must be closely reviewed, both the answer and the supporting narrative. Special attention must be given to all entries that may indicate a negative VPP team finding. If the report indicates that the VPP team possibly misunderstood something they reviewed or did not see all of the supporting information, the VPP Coordinator should clarify the point of the misunderstanding and provide the material to support the request for a revision to the report. One example of this would be if the worksheet contains a finding that the required annual evaluation does not review previous years' recommendations. That determination may have resulted from a number of misunderstandings. The VPP Coordinator may have provided the team with an incomplete draft copy of the last annual evaluation instead of the final complete copy. It is also possible that the provided evaluation was the first formal evaluation and there were no previous formal recommendations. This can be explained in the comments to the report and supported with documentation such as the final complete evaluation report.

Another example of this type of finding may be a negative notation that not all identified hazards are tracked to completion. This may be a result of an incomplete or failed transition from one tracking system to another. OSHA should be satisfied with a confirmation that the tracking system has been corrected and that all items older than 30 days have either been corrected or that interim protections have been provided. An example of the use of interim protection may include the use of respirators while a new ventilation system is installed to control exposure to dust. This situation could also be supported with a copy of a work order and purchase order for the new ventilation system.

Each section of the report is followed by a section summary that allows the team to list 90-day items (hazard correction items), recommendations, and any Merit or one-year conditional goals, if applicable. The site VPP coordinator should provide documentation to OSHA to verify that all 90-day items have been corrected. These are the unsafe conditions and OSHA standards compliance failures found during the evaluation. Without exception, they must all be corrected with verification provided to OSHA. The verification may take the form of picture of the corrective action, a description and example of a program revision, or a statement that necessary training has been completed. That may result in related revisions to the report before it is submitted to the OSHA national office.

The authors suggest that the work site provide a recorder to accompany the OSHA team to record all findings of the team. Those findings should be entered into a report using some form of a table that will list each item with a notation of the person or section responsible for the correction, a due date, an explanation of the corrective action taken, and the correction completion date. The work site should also assign an employee to photograph all of the findings that OSHA points out to them. The verification that all 90-day items have been corrected must be received and accepted by OSHA before any VPP report will be submitted to the OSHA national office for further review and processing. OSHA does allow the use of interim protections so long as the final corrective action is explained.

The worksheet may also contain Merit goals. A Merit goal represents a set of actions that will enable a work site to improve one or more System subelements to a level of quality and effectiveness that would meet the requirements of the VPP Star. This would result when the team detects deficiencies in the safety and health management system, even when physical hazards are not present. The onsite evaluation team must document these deficiencies as goals for correction. Since Merit goals, if assigned, have to be accepted by the applicant, they must be very carefully reviewed. The reviewer must understand the intent of the Merit goal and determine if it is reasonable to expect full and effective implementation within the Merit period. Complete and satisfactory implementation of goals is mandatory for VPP participation. Time frames, interim protection, and methods of achieving goals must be discussed and agreed to by management.

Recommendations are those improvements noted by the onsite evaluation team that are not requirements for VPP participation but that would enhance the effectiveness of the participant's safety and health management system. They differ from the 90-day items in that they are not supported by any OSHA standard and are not required by any statute. One example may be a recommendation to purchase and install an automated external defibrillator (AED), with appropriate employee training. Although there is no specific OSHA standard that requires an AED in the workplace, it has become a standard practice and has been recognized by health care professionals as one of the best responses for heart attack victims. Regardless of the decision, OSHA has to be informed of the decision on the recommendation.

Once the comments are received by OSHA and all corrective actions have been verified, OSHA will make any appropriate revisions to the report and will delete all references to the 90-day items and recommendations. The deletion of 90-day items and recommendations is based on the rational that all 90-day items have been corrected and verified by OSHA and that the recommendations have been responded to. The report will then be referred to the regional administrator for review and then sent to the OSHA national office with the regional administrator's concurrence with the team's recommendation.

The purpose of the subsequent reviews is to ensure that all questions have been answered and that there are no inconsistencies in the report. At the national office the report is reviewed by the VPP staff and if there are no issues, it is sent to the office of the OSHA Assistant Secretary where it is again reviewed. The national office reviews are related mostly to content and consistency. The primary concern is to ensure that all sections have been completed and that the report supports the recommendation. For example, if the recommendation is for Merit, the Merit goal must address a negative finding in the report. Using a weak inspection program as an example, a statement in the report that the entire work site is not inspected quarterly must be supported with a Merit goal that the inspection program be revised to ensure full site inspections at least quarterly. The Merit work site would then be expected to perform full quarterly inspections for at least 12 months to demonstrate completion of the Merit goal.

Once the OSHA national office has completed its review of the report and all questions have been answered, the assistant secretary will approve the work site for participation in the VPP. The Assistant Secretary will sign a letter of congratulations to the work-site manager and notify the OSHA Regional Administrator and the Regional VPP Manager of the decision. All unions representing the work-site employees will also be notified of the decision. The Regional VPP Manager will then notify the local Area Director and will order a VPP flag and a plaque signifying that the work site has been approved as an OSHA VPP participant. The Regional VPP Manager will also prepare a press release that will be submitted to the local newspapers that serve the area in which the work site is located. The work-site management can also prepare their own press releases as well.

CELEBRATE

Then it is time to celebrate. Although OSHA can just mail the plaque and certificate to the work site, they recommend that the participant have a formal presentation ceremony. This provides another excellent opportunity to involve the employees in the VPP process. One of the most successful VPP ceremonies attended by the authors was arranged by the employees of a new VPP Star participant. This occurred at a chemical research and development facility in New Jersey. Although the initial prices of this example may seem exorbitant, it must be noted that the facility was very large with a very large staff. The company originally allocated $25,000.00 for the ceremony. A group of employees proposed to management that they could do the entire ceremony for a total of only $11,000.00. Instead of the original elaborate plan, the employees allocated only $1000.00 for light refreshments for a luncheon presentation. They then selected two local charitable organizations to contribute $5,000.00 each. The organizations were represented at the ceremony, along with the local press and politicians. Not only did the employees realize a great sense of ownership for the very successful ceremony, the company received great press coverage and community relations and saved $14,000.00. This was repeated 4 years later for the second VPP ceremony.

The point is that there is no one formula for a successful ceremony. Factors that have to be considered include the size of the workplace and the funds available. In the history of the VPP there have probably been as many different types of ceremonies as there are VPP sites. Some examples of past VPP ceremonies include mailing the flag and plaque without any fanfare, breakfast meetings with breakfast pizza (yes, pizza with toppings of eggs and sausages) or coffee and cake, plain luncheons such as pizza or sandwiches, more elaborate buffet luncheons, lunch cruises, buffet dinners, and full dinner dances for all employees and their spouses or significant others. The possibilities are endless and the choice is up to the management and associates of the work site given the funds available.

In addition to all local employees, the following should also be invited. Corporate officials of the company, local, state, and federal political representatives, local emergency response organizations such as the fire department and ambulance company, local press, resident contractors, and of course all associates. The OSHA Regional Administrator and Regional OSHA VPP Manager should be invited, as well as the local OSHA Area Director, and the VPP team. The OHSA Regional VPP Manager will notify the OSHA national office about the ceremony and they may attend or send a representative.

The agenda for the ceremony is also not something that is prescribed by the VPP Policies and Procedures Manual. It usually starts after the attendees have had an opportunity to finish their refreshments. The plant manager or the site VPP coordinator should welcome all those present and after an appropriate safety message identify the invited "dignitaries." After a few welcoming words and thanks to all of the employees for their commitment to safety and health and their assistance in achieving the recognition, the dignitaries are invited to speak. Usually OSHA is the last to speak so that they can end the ceremony with the formal presentation. Before OSHA speaks they should be informed of who will receive the flag and plaque. The actual recipients should include both management and hourly employees. Management can be represented by the plant manager and the employees by the safety and health or VPP committee, and the senior union official, if applicable.

At the end of the OSHA presentation the VPP manager may say something like: "Now that the work is done and you have achieved the goal of achieving VPP participation you can all take a deep breath and relax" [after about five seconds the speech continues with]. "Now it is back to work on improving your safety and health management system to the next level."

ANNUAL EVALUATION

Since VPP does not mean Voluntary *Perfection* Programs, OSHA is always expecting VPP work sites to continuously look for ways of improving the safety and health management systems. To track how the VPP sites are performing, each site must complete a full comprehensive annual evaluation of its safety and health management system. The resulting report must be completed in one of the formats recommended by OSHA and must be submitted to the regional VPP manager by February 15 of the following year.

It is important to understand what the annual evaluation is and is not. The methods to effectively and efficiently complete the annual evaluation must also be understood. The annual evaluation is not an assessment of how the safety and health system is working at the time of the evaluation. Although it includes a physical review of the entire work site, it is also not an inspection. One analogy compares the annual evaluation with two types of doctor visits. When someone is sick or hurt, he or she goes to a doctor for a very specific reason. There is a focused complaint and the doctor addresses that complaint with a focused examination and appropriate treatment. For example, a cold may be treated with rest and lots of fluids (and usually chicken soup), and a fracture would be reset and splinted. These visits would not usually include a full-body examination, blood workup, or an EKG.

Along the lines of the VPP annual evaluation, recommended medical practice suggests an annual comprehensive examination by a doctor. During that visit, although there is no specific medical complaint, the doctor performs a full examination of the patient. It usually starts with a question of how the patient has been feeling since the last full examination and if anything has changed. The doctor then examines the patient's entire body and usually prescribes a full blood workup and EKG. The results are then compared to the results of the same tests from the last annual examination. All significant deviations are addressed with additional tests and or treatments. For example, an elevation in the cholesterol level or blood pressure would probably result in a prescription for medication and exercise. This visit also provides the doctor an opportunity to address new medical concerns such as the recent understanding that many in the United States do not have sufficient levels of vitamin D in their blood. Recently, doctors have been recommending that their patients try to increase their vitamin D levels with additional exposure to the sun (of course, the dermatologists disagree) or the use of supplements.

To put this into the perspective of safety and health, let us first consider the example of the patient that is ill or has sustained a fracture. The safety and health equivalent to this would be identifying and correcting an unsafe or unhealthful condition or practice. Should the ventilation system break down, the employees may be exposed to higher levels of harmful agents such as dust or hazardous chemicals. The corrective action would be to remove the employees from the area and repair the ventilation system. This would be the equivalent of the sick patient who is told to rest (removal from the area) and a prescription (repair the ventilation system). During the routine inspection, the inspection team may have observed a broken stair. The correction for this is to reset the stair and repair it. Using the above analogy, this is the equivalent of resetting and splinting the fracture.

The last part of the analogy referred to the doctor reviewing recent medical recommendations and discussing them with patients, as in the need for more vitamin D. The safety and health equivalent of this example is the need to remain on top of all new and revised OSHA standards and regulations and to apply those changes to the programs and practices in the workplace.

The annual evaluation is more similar to the annual visit to the doctor. It is a comprehensive review of the effectiveness of the safety and health management system over the entire past 12 months. In fact, we should emphasize that the implication of doing the evaluation annually is not just once a year but rather every 12 months. It is not sufficient to perform the evaluation in January of one year and then again in December of the following year. That leaves an actual gap of almost 24 months or 2 years between evaluations. That is too long a gap to allow any weaknesses to continue without being identified and corrected. Just imagine the doctor's reaction when you visit after almost 2 years.

Like the OSHA VPP onsite evaluation, the annual evaluation must address all system elements and subelements. In fact, the VPP evaluation process is an effective method to perform the annual evaluation. It should be based on the three-phase process including an audit of the entire work site, a review of all supporting documents, and employee interviews. Before we discuss each of these phases in detail, we will discuss the principles of, and methods to complete, the evaluation.

The basis of the evaluation is to determine how effectively the safety and health management system has been for the period under review. The evaluation should describe how the work site has improved each element and subelement since the previous year and the completion of the previous year's recommendations. It should also describe all deficiencies identified, recommendations for improvement, the person(s) responsible for fulfilling each new recommendation, target dates for their completion, and the data/information reviewed to assess the effectiveness of the subelement.

It is difficult to determine how effective the system is unless there is an understanding of what was found in the previous evaluations. Therefore, it is strongly suggested that any evaluation starts with a review of the previous year's report. To refer to the doctor analogy, the doctor reviews the lab results from the previous year's visit to determine if there have been any changes, for the better or worse. If the previous year's recommendations have not been completed, it is very unlikely that the system is working effectively. Another clear measure of effectiveness is to compare the work-site's injury/illness rates for the last year to those of previous years. Similar is the review of the results of the blood workups for the past year compared to the present results.

ANNUAL EVALUATION METHODS

Before we continue discussing the evaluation process, let us address the method to perform the evaluation. The first instinct would be to assign it to the safety and health manager or an equivalent position. The theory for this is that the safety and health manager is most familiar with the system and how it is supposed to work. While that is true, contrary to the old adage of "physician heal thyself," the better one is "a lawyer who defends himself has a fool for a client." We are not implying that safety and health professionals (including ourselves) are fools. The concern is that safety and health managers may not recognize the weaknesses of the system because they are too close to it. It becomes more of a case of you "can't see the forest for the trees."

Therefore, we strongly recommend that others work with the safety and health manager to perform the annual evaluation. This is also an excellent opportunity to involve more employees to participate in the safety and health management system. One method that has been found to be very effective is to organize several focus teams. These teams would be assigned one or more system elements and subelements to evaluate. Of course, this method raises some issues such as the subject knowledge of the expectations of the assigned elements and the specific regulatory requirements. This could be addressed by providing orientation and training to the team members and assigning a subject matter expert to the individual teams. Care must be taken to avoid using an element owner to avoid any sense of loss of objectivity. Of course, the element owner would provide input into the evaluation and assist the team. The training would also have to include instructions to the team about the format of the annual evaluation. The safety and health manager would be responsible for ensuring that all training is provided and the evaluation proceeds according to schedule.

Another method that has been found to be effective is to select a management team to complete the entire evaluation by reviewing all elements and subelements.

The benefit of this method is that the team members may be more familiar with the process and the details of the safety and health management system. One concern with this is the reduction in employee involvement and the time required for each of the team members. By reducing the number of employees involved in the process, it adds more responsibility and time requirements to those that are on the team. Another concern with this method is that it limits the input of additional points of view.

A third method is for the safety and health manager to complete the annual evaluation. Clearly, this method benefits in that it limits the resources necessary, and the safety and health manager would clearly be familiar with the process, the system, the programs and procedures, and the applicable regulations and standards. However, this method has the issue of a potential loss of objectivity since the reviewer is most likely the one who wrote the programs and procedures. That position is also likely to be the one that is responsible for the continuing reviews.

Some corporations provide an evaluation service to each of its operating facilities. These are performed with some regularity based on a predetermined schedule. This is a very effective method in that it does not require much in the way of facility resources, other than providing support to the corporate auditors during their evaluation. The auditors are also usually specifically trained in the evaluation process, as well as the requirements and the applicable programs and procedures. The auditors are also usually safety and health professionals. Some corporations provide this type of service in a slightly different fashion. Instead of using corporate staff safety and health professionals, the evaluation teams are composed of safety and health professionals from other similar corporate facilities. The strongest benefit of this method is that those performing the evaluation are intimately familiar with the process and the elements. It also provides the benefit of the auditors being able to benchmark their own activities with those of the facilities they are evaluating. It also encourages a sharing of best practices.

Yet one more method is to use an outside safety and health organization to perform the evaluation. There are third-party companies that specialize in performing safety and health system evaluations that are tailored to the VPP requirements. This method usually provides the most objective evaluation.

Regardless of the method chosen to perform the annual evaluation, the evaluation process should follow the one recommended by OSHA in the VPP Policies and Procedures Manual. That process results in a report that contains the following sections: the general information about the work site, a review of recommendations of the previous years' annual evaluation, a review of the OSHA logs, a table of injuries and illnesses, a description of all significant changes to each of the VPP elements, an assessment of each VPP element, appropriate recommendations to improve those elements that were recognized as having some deficiencies, a description of how Merit or one-year conditional goals have been addressed, all significant safety- and health-related success stories, and a description of any VPP activities that may be considered as supporting the VPP in the community or with other companies. Each of these sections will be discussed in greater detail.

It is also suggested that the self-evaluation follow the procedures used by OSHA during their VPP onsite evaluations. That is using a three-part process of employee interviews, document reviews, and physical observations of the entire facility. There are several ways that the employee interviews may be conducted. Regardless of the method chosen, care must be taken to avoid any situation that may inhibit employees from expressing their true feelings. For example, it would be inappropriate for a supervisor to interview a direct report and then ask: "How am I doing in demonstrating support for your safety and health." One way to avoid this situation would be for peers to interview peers. Another suggestion that is often used to measure the employees perceptions of management's leadership and commitment to their safety and health and their active involvement in the system is the use of anonymous surveys. Although this avoids the concern for intimidation, many companies do not have the technical expertise or the funds available for hiring an outside consultant to perform the survey. Another problem with the survey is that many surveys do not result in an adequate return of the survey forms. Regardless of the method used to obtain feedback from the employees, the annual evaluation would not be considered complete or effective without it.

The document review during the annual evaluation is not so much an audit to confirm that the programs meet or exceed the requirements of each of the applicable OSHA standards as it is an effort to determine its effectiveness to ensure a safe and healthful workplace. Clearly, one must consider the regulatory requirements to ensure compliance with them, but it is the application of the individual programs that must be addressed in the annual evaluation. Consideration must also be given to reviewing the effectiveness of those nonregulatory VPP requirements, such as monthly inspections and hazard tracking. We will present two examples of considerations for reviewing hazard prevention programs and two examples of considerations for reviewing other nonregulatory VPP requirements:

A. Regulatory Examples

1. *Hazard Communication Program* There must be a site-specific written program that is reviewed to ensure that all elements have been included and that the program still meets the requirements of the facility. A typical non-VPP work site may review this document and be satisfied with it so long as the program includes all of the standard's requirements, such as labels, MSDSs, and training. A VPP work site would review the document to ensure that all parts of it have been fully implemented and are working effectively. This review would consist of a workplace audit to inspect all chemical containers for proper labeling; the MSDSs would be reviewed to filter out all unused chemicals and add all recently purchased chemicals; the training would be reviewed to identify those employees that may not have received the training, as well as to determine the effectiveness of the training by tests or other methods. In addition, the review should address how effectively the employees responded to any recent spills or other exposures to the covered chemicals.
2. *Confined-Space Program* A typical non-VPP work site may limit its review to just ensuring that there is a written program that meets all of the standard's

requirements such as training, identification, entry procedures, and rescue plans. The VPP work site would expand this review to include a review of previous confined-space entry permits (looking at their completeness, calibration and testing notes, signatures, and rescue provisions). The training, as with the hazard communication program, must be evaluated for its effectiveness and currency. The rescue provisions would be reviewed to confirm that training has been completed and that routine drills have been conducted and critiqued. Deficiencies should result in revisions to the program.

B. Nonregulatory VPP Requirements

1. *Employee Participation* Other than the few standards that actually require employee involvement (e.g., Process Safety Management), the VPP is the only other requirement that OSHA has for employee involvement. The VPP requires that employees be provided at least three means of being actively and meaningfully involved in the safety and health management system. To avoid any confusion, the requirement is not that all employees become involved, simply that they are provided the opportunity and encouraged to do so. One measure of effectiveness of this requirement would be to survey the employees to determine how each of them is involved. The survey could also be used to determine why they may not be. Other sources of determining the level of employee involvement would include reviewing their individual participation in such activities as inspections, job hazard analyses, membership on committees and safety-related teams, and suggestions submitted. The annual review of the level of involvement would then be compared to the level of involvement of the previous year.
2. *Job Hazard Analyses (JHA)* As in the above example, there are only a few OSHA standards that specifically require a hazard analysis (e.g., Process Safety Management). However, the VPP expects that all hazardous routine tasks be evaluated so that preventive actions can be developed to control the identified hazards. As part of the annual evaluation, all JHAs must be reviewed to determine their continued effectiveness and be revised to recognize newer technology or chemicals. Effectiveness would include considerations such as the percentage of JHAs reviewed, the percentage of JHAs revised, and the percentage of employees participating in the JHA reviews. The evaluation would also review the training the employees have received on the JHAs.

The third leg of the evaluation process is the direct observation of the workplace. This is not a historical review but rather a picture in time. In fact, it is actually an inspection of the entire work site, and it is critical to the annual evaluation process. The direct observation of the workplace is one of the primary methods of determining the effectiveness of the safety and health elements, procedures, policies, and programs. For example, finding numerous hazards may indicate that one or more of the following elements are deficient: inspections, training, maintenance, tracking, and the like. The observation process may also disclose unsafe employee practices such as improper use of PPE or work practice shortcuts. The observations made during this inspection must result in the prompt correction of all hazards and unsafe practices. It should also result in more in-depth evaluations of the programs and procedures that should have prevented them from occurring in the first place.

RECOMMENDATIONS

Since OSHA works under the principle that the VPP does not mean Voluntary *Perfection* Programs, they do not expect perfection in the safety and health management systems of the VPP work sites and applicants, and the underlying principle of the VPP is one of continuous improvement, an evaluation that does not contain any recommendations for improvement may be viewed with some skepticism. That is not to say that it is impossible that an evaluation would result in no recommendations. It is just that OSHA's experience in the VPP has shown that there is almost always room for improvement. This is one reason that we suggest that instead of the owner of the safety and health management system, others be involved in the review.

Recommendations may address either identified areas of improvement or weaknesses in any one or more VPP elements. One example of an area of improvement may be a recognition that employees are not as involved in safety and health activities as would be expected because there is too much focus on production and time is not allowed. This may result in a recommendation to increase employee involvement through the development of additional opportunities and a focus on safety and health rather than production. Another example of an area of improvement may be a determination that the training program has been using the same training videos for the last several years. This usually results in a lack of interest and poor training. Developing new training programs, with employee input, could be a recommendation for this concern. Success in each of these recommendations can be easily measured by surveying employee involvement and the use of tests to determine the effectiveness of the new training programs. Employee participation during the training is also a strong indicator of effectiveness.

A last example of an observed area of improvement may be disclosed weaknesses in the inspection program. As part of the annual evaluation, a full detailed workplace audit or inspection must be completed. This is best done by a trained independent source since they will represent a new set of eyes. While it is expected that third-party audits will most likely find new and more unsafe conditions, it is the nature and number of those findings that may be a cause for concern. It is typical to find a few unlabeled chemical spray bottles, especially in the maintenance department. It is also typical to find a limited number of other unsafe conditions such as dirty eyewashes stations, expired fire extinguishers, emergency exits blocked with empty boxes, and bench grinders out of allowed tolerances. As long as these situations are limited, and correction is completed promptly, there is not usually much cause for concern. If these examples are found to be endemic, then it is assumed that there is a deficiency in the inspection program. That deficiency may be related to an inadequate inspection schedule, a lack of follow-up on observed hazards, or more likely a need for more work-site-specific hazard recognition training for those performing the inspections. A closer review of the inspection program would assist in determining the true weakness and facilitate the development of the recommendation for improvement.

Recommendations such as those described above would be included in the report. Responsibility for action on each of the recommendations must be identified and there should be a specified timeline for its completion. All of the recommendations in the report should be entered into some form of tracking system and must be tracked on a continuing basis with occasional status reports.

ANNUAL EVALUATION FORMAT

The OSHA Voluntary Protection Programs Policies and Procedures Manual contains a recommended format for the annual evaluation. Several OSHA Regional VPP Managers have developed their own formats and instructions for the annual evaluation. They also usually send a notice out to all VPP participants to remind them of the requirement to submit the reports no later than February 15. An example of the letter and instruction that are sent by OSHA about the annual evaluation are presented in Exhibit 9.1.

The annual evaluation usually consists of a least three sections. These sections include:

- Section I: Summary Sheet
- Section II: Injury/Illness Rates
- Section III: Narrative Evaluation of Safety and Health Management System
  - Previous year's goals and recommendations status report
  - Current year's goals and any recommendations that were developed from your annual self-evaluation
  - Summary chart of Merit or one-year conditional goals
  - Other information
  - Significant changes or Events
  - Narrative evaluation of safety and health management system (for each subelement)

The annual evaluation format from the VPP Policies and Procedures Manual is presented in Exhibit 9.2. The authors have added explanatory notes to the format. A sample of an annual evaluation is presented in Exhibit 9.3.

Exhibit 9.1 Example Letter and Instruction Sent by OSHA about the Annual Evaluation

U.S. Department of Labor

Occupational Safety and Health Administration

Regional Office

September 9, 2009

Letter To All Regional Voluntary Protection Programs (VPP) Participants

Dear VPP Participant:

This is a courtesy reminder that all VPP participant sites are required by the Federal Register to submit the following by February 15: your site's injury and illness data and the annual evaluation of your safety and health management system for the previous year. This includes all combined injury and illness cases (the combined total of columns H, I and J from the OSHA 300 logs) and cases involving Days Away, Transferred or Restricted (columns H and I). In addition, you must submit the total hours worked, average annual employment, average contractor employment and any explanatory information. This information MUST be included on the attached table!

Please include any success stories on the last page. This showcases your successes and helps us to respond expeditiously to National Office inquiries. We need an explanation for any increases in your rates and a descriptive plan of action for reduction. I have included a form that you should use to submit your injury/illness data. This form will also be used to update our records. The same information must be submitted for all applicable contractors' employees on the site.

The report must be received by me no later than Feb. 15, 20XY.

If you have questions, please feel free to contact me.

Sincerely,

Regional VPP Manager

Attachment

Annual Evaluation of Site Safety and Health Management System

Discussion:

Participation in the VPP includes the requirement that each site conduct an annual evaluation of their safety and health management system. The evaluation must:

A. Provide for written narrative reports with recommendations for improvements and documentation of follow-up action. To be effective, the evaluation must provide for timely correction of any areas in need of improvement, and include assignment of responsibility for corrective action.

B. The evaluation must assess the effectiveness of all required VPP elements and any other of the site's safety and health management system.

C. The evaluation may be conducted by competent corporate or site personnel or by competent private sector third parties who are trained and/or experienced in performing such evaluations. Annual Evaluations combined with other corporate audits are acceptable only if they cover all of the elements discussed in this format.

The evaluation should be conducted annually at approximately the same time each year. It should follow the format outlined in the VPP requirements described below. The effectiveness of each element and sub-element must be assessed briefly in narrative form. The recommendations for improvement should follow each narrative. The recommendations should be summarized again at the end of the evaluation along with the assignment of responsibility for completing the recommended improvements and the target date for correction.

Your evaluation should be conducted similar to OSHA's onsite review. It should include a review of your written program, a walk-through of your workplace, and interviews with employees. During the evaluation you should be answering the questions presented in CSP 01-03-002 (formally TED 8.1), Appendix D.

This is the document providing the guidelines to the OSHA team to determine a site's eligibility for the VPP, or to remain in the VPP as a Star, Demonstration, or Merit site. Also ask yourself these questions:

Is it comprehensive?

Is it operating effectively?

What improvements can be made to make it more effective?

The evaluation should follow the following format or the one described in Appendix D of the CSP.

REMEMBER: A self-evaluation is NOT an inspection of the worksite; it is a critical review of ALL of the elements of the safety and health management system. An evaluation that is merely a workplace inspection with a brief report pointing out hazards or saying that everything is okay is inadequate for purposes of VPP.

For a more comprehensive description of the VPP requirements refer to the federal register for VPP and to the CSP directive.

The purpose of providing OSHA with a copy of your annual evaluation is to allow us to periodically review your system's condition without visiting your site to determine the need for a site visit by OSHA. We are now conducting initial reevaluation visits three and one-half-years after approval, and subsequent evaluations four and one-half years after the previous one.

Evaluation of Elements and Sub-Elements:

In assessing your system use the following format to address each element and sub-element addressed in sections IV-VII of this format:

a. Provide a summary description of the sub-element. It is a very good idea to use the questions provided in Appendix E of the CSP directive, "Onsite Evaluation Report Format" to provide additional guidance.

b. In narrative form, assess the effectiveness of each of the elements and sub-elements listed on the next page. Focus should be on the improvements made since the previous year and completion of the previous year's recommendations.

c. Include in each element any deficiencies identified and recommendations for improvement.

d. For each recommendation for improvement assign a person(s) to be responsible for completing each recommendation and meeting target dates (see section "e").

e. Assign target dates for completion of recommendation.

f. The data/information reviewed to assess the effectiveness of the sub-element.

10

VPP REAPPROVAL

CONTINUOUS IMPROVEMENT

The VPP stands for Voluntary Protection Programs; it does not stand for Voluntary Perfection Programs. OSHA does not expect VPP work sites to be perfect. Instead the VPP was originally developed on the unofficial principle that to accept perfection is to eliminate the challenge to continuously improve the safety and health management system. Continuous improvement has been a cornerstone of the VPP since its inception. In the revisions to the VPP published in January, 2009, OSHA has made the principle of continuous improvement an explicit principle of the VPP. The principle of continuous improvement is one of the reasons that VPP work sites continue to see their recordable rates decline and the employees become more involved in the safety and health activities at the workplace. It is also a primary contributor to the fact that VPP work sites lead most other work sites in injury prevention and reduction.

Continuous improvement is also one of the reasons why employees at VPP work sites incur fewer off-the-job injuries than non-VPP work sites. The focus on off-the-job injuries is relatively new, and VPP work sites are leading their industries in recognizing the importance of providing employees the tools to protect themselves and their families. Through the principle of continuous improvement these work sites have realized that many workdays are lost because employees who are injured or become ill during their non-work-related activities cannot report for work. Employees may also be prevented from reporting to work to care for other family members that may have been injured during a non-work-related activity. To prevent

Preparing for OSHA's Voluntary Protection Programs. By Brian T. Bennett and Norman R. Deitch
Copyright © 2010 John Wiley & Sons, Inc.

non-work-related incidents, these work sites have developed programs that include safety and health training in off-the-job activities and have provided employees with appropriate PPE for use away from the job.

The alternative to continuous improvement is complacency. Complacency is probably one of the greatest dangers to the safety and health of workers. The principle of complacency is one of accepting what has happened in the past. It represents satisfaction or contentment with the existing safety and health management system. Since there is satisfaction with history, there will be no challenge to evaluate what was done in the past to ensure the safety and health of the employees and opportunities for improvements will be lost. Many VPP work sites that have been in the VPP for many years are now falling into the complacency trap. The biggest challenge facing VPP work sites that have been in the program for 10 or more years is complacency. These work sites will need to be creative to sustain employee involvement and continue their pursuit of continuous improvement.

To evaluate the commitment of VPP workplaces to the principle of continuous improvement, OSHA has included in both the VPP *Federal Register* and the VPP Policies and Procedures Manual several requirements. These include:

- Annual critical self-evaluations of the safety and health management system with recommendations to address identified opportunities for improvement submitted to OSHA each year by February 15.
- Annual submission to OSHA of tables of injury/illness rates for the most recent past 4 calendar years by February 15 of each year and investigation and explanation of upward trends of injury and illness rates.
- Investigations of employee complaints submitted to OSHA.
- Investigations of significant safety and health events at the work site.
- Routine onsite evaluations:
  - Star sites are reevaluated 30–42 months after their initial approval.
  - Star sites are reevaluated within 60 months after the second and subsequent reevaluations.
  - One-year conditional Star sites are reevaluated within 15 months (90 days plus 1 year's experience operating at Star level) after the participant was placed on conditional status.
  - Merit sites are evaluated 18–24 months after their initial approval and at the end of the Merit term of approval.
  - Star Demonstration participants are reevaluated every 12–18 months after initial approval.
- Nonroutine onsite evaluations:
  - Either full or focused onsite evaluations may be conducted earlier than normal scheduling requirements when:
    - Significant changes have occurred in management, processes, or products that may require evaluation to ensure the participant is maintaining a VPP quality safety and health management system.

- OSHA has learned of significant problems, such as increasing injury and illness rates, serious deficiencies described in the participant's annual evaluation of its safety and health management system, or deficiencies discovered through OSHA enforcement activity resulting from an employee complaint, fatality, catastrophe, or significant event.
- An onsite evaluation may be conducted earlier than is required when requested by a participant.

ANNUAL SELF-EVALUATION

To ensure that the VPP work sites are fulfilling their obligations to improve their safety and health management systems, they must complete a comprehensive self-evaluation each year and submit a copy of that report to OSHA by February 15 of the following year. The self-evaluation is discussed in detail in Chapter 9. Since OSHA does not expect perfect safety and health management systems, they expect that the report will contain at least some recommendations to improve one or more of the elements of the VPP. To identify those opportunities requires a critical introspective look at what was done during the past year with a view toward how to make the element even more effective. Reports that do not contain at least some recommendations for improvement may be looked at with some skepticism by OSHA. The belief is that there is always something that can be done to improve the safety and health management system. One example of such an improvement may be a workplace that considers the level of employee involvement very high. The reality may be that only 40% of the employees are actually actively involved. The challenge would be to increase that percentage, and the recommendations may include new methods of encouragement for the employees to become more directly involved.

During the routine VPP onsite evaluations, OSHA expects to hear what the VPP participant has done since the previous evaluation to improve the safety and health management system. That information would be presented to the OSHA onsite evaluation team during the opening conference.

Examples of improvements to the safety and health management system may include:

- Restructuring of the safety committee to include more off-shift employees
- Revision of the inspection program to include employees that work in other areas
- Development of a subcommittee for the administrative staff
- The addition of new training programs for all employees based on employee suggestions
- Expansion of the salaried employees' annual performance evaluation process to include more measureable safety and health metrics and a 360 review by the direct report employees
- Completion of a safety and health performance survey by the employees

- Providing a new emphasis program to address off-the-job safety
- Installation of an automated electronic defibrillator with supporting employee training
- Initiation of a new employee-run medical emergency response team

Those improvements may be demonstrated by:

- Increased employee involvement
- Reductions in measureable negative safety and health indicators such as injuries, illnesses, first-aid cases, and workers' compensation costs
- Increases in positive safety and health measures such as near-miss reports, reduced backlogs of safety and health work orders, increased safety and health training
- Increases in productivity

Specific examples of recommendations for the improvement of a safety and health management system, VPP Merit goals, and one-year conditional goals are provided in Exhibit 10.1 (on attached CD). Merit Goals, one-year conditional goals, and recommendations can be considered interchangeable given similar but different circumstances. These have been excerpted from numerous OSHA VPP onsite evaluation reports and represent deficiencies that the workplace should have picked up but for whatever reason they were overlooked. It must not be implied that the VPP participant or applicant was not effective in their self-assessment. The examples on Exhibit 10.1 are a compilation of reports for different workplaces and represents a period of reports from several years. The list is provided as an example of the type of improvements OSHA is looking for and expects to find. The list does not contain those requirements that are needed for compliance with OSHA standards. It is broken down by the elements of a safety and health management system. The authors would like to thank Bill Klingbeil, the OSHA Region 6 VPP Manager for compiling this information.

ANNUAL SUBMISSION OF TABLE OF OSHA RECORDABLE RATES

Similar to the table of OSHA recordable incidents, the annual evaluation must include a table that compares the information for the most recent 4 years. The information must be submitted to OSHA no later than February 15 of the following year. The deadline for the submission may be extended by up to 45 days if requested by the participant. Failure to submit it after the 45-day extension may result in a request sent to the participant to withdraw from the VPP.

Table 10.1 includes the calculation of two 3-year rates; one for the most recent 3 years and one for the previous 3 years. That enables the work site and OSHA to compare the 3-year trend.

TABLE 10.1 Annual Evaluation Table of Recordable Injuries and Illnesses

Year	Hours Worked	Total Cases	Total Case Incident Rate (TCIR)**	Days Away/ Restricted/ Transferred Cases	Days Away/ Restricted/ Transferred Incident Rate (DART)
2005	300,000	6	4.0	2	1.3
2006	300,000	3	4.0	2	1.3
2007	300,000	3	2.0	1	0.7
2008	300,000	1	0.7	0	0.0
Total for (2005–2007)	900,000		2.7		1.1
Total for (2006–2008)	900,000		1.6		0.7
Enter your NAICS code here					
____XXXXXX____			1.2		0.9
2007 BLS Average for NAICS***					
Percent Above/Below BLS Rate:			33		22

From the data in Table 10.1 it can be seen that the most recent 3-year rates (2006–2008) are substantially lower than the previous 3-year rates (2005–2007). In reviewing this table OSHA would be pleased to note that the 3-year TCIR rate was reduced from 2.7 to 1.6, a 40% reduction of total recordable incidents. The rates for days away from work cases dropped 36% from 1.1 to 0.7.

Should the 3-year rates increase from one report period to another, OSHA will contact the work site to determine the reason and to inquire what steps the participant has planned to address the increase. In cases where the most recent 3-year rates exceed the BLS rates for any of the most recent 3 years, the participant will receive a notice from OSHA that they have to prepare a formal rate reduction plan. The rate reduction plan must detail the steps that the participant will initiate to reduce the rates to below the BLS rates within 2 years. The 2-year rate reduction plan must be approved by OSHA as being appropriate and realistic.

INVESTIGATIONS OF EMPLOYEE COMPLAINTS SUBMITTED TO OSHA

Participation in the VPP does not eliminate the rights that are afforded to all employees by the OSH Act. Those rights include the ability to file a complaint with OSHA without fear of reprisal or discrimination. There are generally two types of complaints that OSHA can receive. When either one is investigated by OSHA, it will be done with the strictest confidence to protect the complainant. One is the informal complaint that may be submitted by phone or an unsigned letter. The informal complaint may be investigated by OSHA by simply calling the work site and inquiring about its validity. The more serious complaint is referred to as a formal complaint that must be signed by the complainant. When a formal complaint is received against a VPP participant, they are investigated in the same manner as for any other work site. The only difference is that notification of the complaint will be given to the regional VPP Manager. The result of the complaint inspection will also be provided to the regional OSHA VPP Manager so a determination can be made if the issue was simply one of a compliance nature or if it is reflective of a weakness in any element of the safety and health management system. Perceived weaknesses will be investigated after the complaint inspections have been completed.

INVESTIGATIONS OF SIGNIFICANT SAFETY AND HEALTH EVENTS AT THE WORK SITE

In addition to complaints, OSHA may receive a referral from other agencies or they may be notified of a fatality, catastrophe, or other event requiring OSHA enforcement activity at a VPP work site. In such instances the OSHA Area Director must initiate an inspection following normal OSHA enforcement procedures. The VPP Regional Manager will be informed of the action, and any ongoing VPP activity at the work site will immediately cease, with the inspection taking precedence. All VPP activity will be held in abeyance until the inspection case is formally closed.

Upon completion of the inspection a determination will be made by the OSHA Regional VPP Manager if the cause of the action was the result of an obvious systemic error in the participant's safety and health management system. The review will also determine the participant's level of cooperation with the investigation, the type of any OSHA citation issued (willful or serious), confirmation that all cited hazards were abated, and that all VPP elements continue to be in place. Although this may be done by phone, it is usually done by a visit to the participant by either the OSHA Regional VPP Manager or the local OSHA compliance assistance specialist (CAS). This visit is not in the form of a VPP onsite evaluation and is relatively informal. It is very focused on the causes of the incident that led to the OSHA inspection. The purpose of the inquiry, either by phone or in person, is to:

1. Obtain assurances that management and unions (if applicable) remain committed to VPP.
2. Note any improvements in the participant's systems, policies, procedures, and/or hazard controls.
3. Determine whether the participant remains qualified for VPP participation.

OSHA inspections that result in an upheld citation for a willful violation do not automatically result in the termination of a participant's participation. However, the OSHA VPP Manager must closely review the case. If it is obvious that the participant no longer meets the requirements of the VPP, then no onsite will be performed and procedures for withdrawal or termination will be initiated.

Those procedures begin with a request from the OSHA regional administrator to the participant to withdraw from participation in the VPP. Should the participant volunteer to withdraw, they will send a letter to the VPP Manager requesting the withdrawal, without prejudice. If the site does not request a withdrawal, the site will be removed from the program, and that withdrawal is with prejudice should the site attempt to reapply in the future.

ROUTINE ONSITE EVALUATIONS

One of the primary benefits of the VPP is the requirement for routine onsite evaluations by OSHA of the participant's safety and health management system. This provision embraces the concept of continuous improvement by requiring work sites to perform an annual critical assessment of their safety and health management system.

OSHA will perform these routine evaluations following the schedule detailed below:

- Star sites are reevaluated 30–42 months after their initial approval.
- Star sites are reevaluated within 60 months after the second and subsequent reevaluations.
- One-year conditional Star sites are reevaluated within 15 months (90 days plus 1 year's experience operating at Star level) after the participant was placed on conditional status.
- Merit sites are evaluated 18–24 months after their initial approval and at the end of the Merit term of approval.
- Star Demonstration participants are reevaluated every 12–18 months after initial approval.

The first onsite evaluation following the initial approval follows the same procedure and format as the initial onsite evaluation as described in Chapter 9. There is no requirement to submit what some have referred to as a reevaluation application. The OSHA VPP Manager will call you to schedule a mutually convenient date to conduct the reevaluation.

The participant should be prepared to describe the improvements to the safety and health management systems since the initial VPP onsite evaluation and the results of those improvements.

Reevaluations for Star participants, subsequent to the first reevaluation, may follow a more streamlined procedure called the Compressed Reapproval Process to Recognize Sustained Excellence (CRP). To be eligible for the CRP, the participant must meet the following requirements:

1. Be in compliance with all assurances agreed to in the VPP application.
2. The most recent annual evaluation was complete and demonstrated VPP quality safety and health excellence.

3. Be in good standing at the Star level, that is, not under a Star conditional status or a rate reduction plan.
4. Have experienced no work-related fatalities or catastrophes since the most recent VPP onsite evaluation.
5. Have not received willful, repeat, or high gravity serious citations since the most recent VPP onsite evaluation.
6. The most recent 3-year injury and illness rates (TCIR and DART) still meet Star requirements. However, the OSHA VPP manager may determine that irregularities within rates that otherwise meet this requirement (e.g., rates that trend up) warrant a comprehensive onsite evaluation.
7. The OSHA VPP manager determines that the participant's OSHA complaint history and findings since its most recent VPP onsite evaluation do not indicate the need for a comprehensive onsite evaluation. In making this determination, the OSHA VPP Manager will consider the participant's size, complexity, and work culture.
8. The OSHA VPP Manager determines that the changes to management, ownership, or bargaining unit status that have been reported to OSHA do not warrant a comprehensive onsite evaluation.

Should the OSHA VPP Manager determine that the CRP is appropriate, the participant will be notified of the decision and what to expect. The participant will also be informed that the CRP evaluation may be expanded to a full comprehensive evaluation depending on the findings of the onsite evaluation team. Some reasons that the evaluation may be expanded include:

- Workplace conditions are not up to the standard expected at Star work sites.
- Employees have commented that management commitment to the VPP has dropped since the last evaluation as demonstrated by nonattendance at VPPPA conferences.
- Programs have not been reviewed regularly.
- Training has been missed by many employees.
- The safety committee has been disbanded with no suitable alternative for employee involvement.

The CRP follows the same process as the comprehensive VPP onsite evaluation. The differences are in the scope of the process and the final written report of the team's findings. The CRP evaluation will include a walkthrough of the entire work site that pays special attention to any changes in equipment, process flow, and/or operating procedures. It will also include a review of the safety and health documents including the safety and health programs. The document review will focus on new and modified policies, programs, and procedures and on highly hazardous processes. Although there will be interviews by the VPP evaluation team, they will be mostly informal and primarily with managers and supervisors and some employees.

For participants who produce or use highly hazardous chemicals, as defined in OSHA's Process Safety Management (PSM) regulations, a process safety review must be conducted by a team member qualified to evaluate PSM in accordance with VPP procedures. The findings of this review must be included on the onsite evaluation worksheet.

The compressed reapproval process report is similar to that of a comprehensive evaluation with the difference that only those areas that are considered minimum requirements for participation in the VPP are included. Exhibit 10.2 is the report worksheet for the CRP.

NONROUTINE ONSITE EVALUATIONS

There are a few reasons that the regional OSHA VPP Manager may decide to initiate either a full or focused onsite evaluation earlier than normal scheduling requirements. These may be related to OSHA enforcement actions, significant organizational or other changes to the participant, or a request received from the management or union of the participant. Following are some examples of what may lead to a nonroutine onsite evaluation:

- OSHA inspection activity
- When an OSHA enforcement inspection raises concerns about the potential for major failures in the participant's safety and health management system, this indicates that not all VPP elements are in place or are not effective, or a fatality or multiple fatalities have occurred. The evaluation may be initiated to:
  - Review all safety and health management system elements.
  - Obtain assurances that management and unions (if applicable) remain committed to VPP.
  - Determine if the participant remains qualified for VPP participation.
- Significant change to the participant:
  - When changes have occurred in management, processes, or products that may require evaluation to ensure the participant is maintaining a VPP quality safety and health management system. This would not usually result in the simple change of a plant manager. It would more likely result from the change of several senior managers at the participant.
  - OSHA has learned of significant problems, such as increasing injury and illness rates, serious deficiencies described in the participant's annual evaluation of its safety and health management system.
- An onsite evaluation may be conducted earlier than is required when requested by a participant or the union. The request may be to address concerns from either the participant or the union that the safety and health management system may have deteriorated with no measureable deficiencies. For example, the union may have a perception that the commitment of management to the VPP has

fallen as evidenced by their limiting union participation in the Special Government Employee Program and their attendance at the VPPPA conferences. Management may request an accelerated onsite evaluation because they have initiated new programs that they would like OSHA to review as soon as possible.

Exhibit 10.1 Examples of Recommendations for Improvement of Safety and Health Management System, VPP Merit Goals, and One-Year Conditional Goals

Exhibit 10.2 Annual Report Worksheet for Compressed Review Process

Section I: Management Leadership & Employee Involvement	Yes or No	How Assessed: Interview	Observation	Doc Review
A. Written Safety & Health Management System				
A3. *Is the written safety and health management system at least minimally effective to address the scope and complexity of the hazards at the site? (Smaller, less complex sites require a less complex system.) If not, please explain.*				
•				
B. Management Commitment & Leadership				
B1. *Does management overall demonstrate at least minimally effective, visible leadership with respect to the safety and health program (considering FRN items F5 A-H)? Provide examples.*				
•				
D. Authority and Line Accountability				
D1. *Does top management accept ultimate responsibility for safety and health in the organization? (Top management acknowledges ultimate responsibility even if some safety and health functions are delegated to others.) If not, please explain.*				
•				
D3. *Do the individuals assigned responsibility for safety and health have the authority to ensure that hazards are corrected or necessary changes to the safety and health management system are made? If not, please explain.*				
•				
D5. *Are adequate resources (equipment, budget, or experts) dedicated to ensuring workplace safety and health? Provide examples.*				
•				

E. Contract Workers				
E5. *Does the site's contractor program cover the prompt correction and control of hazards in the event that the contractor fails to correct or control such hazards? Provide examples.*				
•				
E11. *Is the contract oversight minimally effective for the nature of the site? (Inadequate oversight is indicated by significant hazards created by the contractor, employees exposed to hazards, or a lack of host audits.) If not, please explain.*				
•				
F. Employee Involvement				
F3. *Do employees support the site's participation in the VPP Process?*				
•				
F4. *Do employees feel free to participate in the safety and health management system without fear of discrimination or reprisal? If so, please explain.*				
•				

Section I: Management Leadership & Employee Involvement
Conditional Goals *(Include cross reference to section, subsection, and question, e.g., I.B2)*
1.

90-Day Items *(Delete this section for final transmittal to National Office)*
1.

Best Practices
1.

Comments including Recommendations (*Optional*)
1.

Documents Referenced, Programs Reviewed (*Optional*)
1.

Section II: Worksite Analysis	Yes or No	How Assessed: Interview	Observation	Doc Review
A. Baseline Hazard Analysis				
A1. *Has the site been at least minimally effective at identifying and documenting the common safety and health hazards associated with the site (such as those found in OSHA regulations, building standards, etc., and for which existing controls are well known)? If not, please explain.*				
•				
C. Hazard Analysis of Routine Activities				
C1. *Is there at least a minimally effective hazard analysis system in place for routine operations and activities?*				
•				

D. Routine Inspections				
D1. *Does the site have a minimally effective system for performing safety and health inspections (i.e., a minimally effective system identifies hazards associated with normal operations)? If not, please explain.* •				
E. Hazard Reporting				
E3. *Is there a minimally effective means for employees to report hazards and have them addressed? If not, please explain.* •				
F. Hazard Tracking				
F4. *Does a minimally effective tracking system exist that results in hazards being controlled? If not, please explain.* •				
G. Accident/Incident Investigations				
G1. *Is there a minimally effective system for conducting accident/incident investigations, including near-misses? If not, please explain.* •				
I. Trend Analysis				
I1. *Does the site have a minimally effective means for identifying and assessing trends?* •				

Section II: Worksite Analysis
Conditional Goals (*Include cross reference to section, subsection, and question, e.g., II.B2*)
1.

90-Day Items (*Delete this section for final transmittal to National Office*)
1.

Best Practices
1.

Comments including Recommendations (*Optional*)
1.

Documents Referenced, Programs Reviewed (*Optional*)
1.

Section III: Hazard Prevention and Control	Yes or No	How Assessed: Interview	Observation	Doc Review

A. Hazard Prevention and Control				
A1. *Does the site select at least minimally effective controls to prevent exposing employees to hazards?*				
•				
A10. *Does the site have minimally effective written procedures for emergencies (TED 3-16 3h)?*				
•				
A18. *Is the site covered by the Process Safety Management standard (29 CFR 1910.119)? If yes, please answer questions A19-A21 below. Additionally, please complete either onsite evaluation supplement A or B, and onsite evaluation supplement C. If not, skip to section B.*				
•				
A19. *Which chemicals that trigger the Process Safety Management (PSM) standard are present?*				
•				
A20. *Which process(es) were followed from beginning to end and used to verify answers to the questions asked in the PSM application supplement, the PSM Questionnaire, and/or the Dynamic Inspection Priority Lists?*				
•				

A21. *Verify that contractor employees who perform maintenance, repair, turnaround, major renovation or specialty work on or adjacent to a covered process have received adequate training and demonstrate appropriate knowledge of hazards associated with PSM, such as non-routine tasks, process hazards, hot work, emergency evacuation procedures, etc. Please explain.*

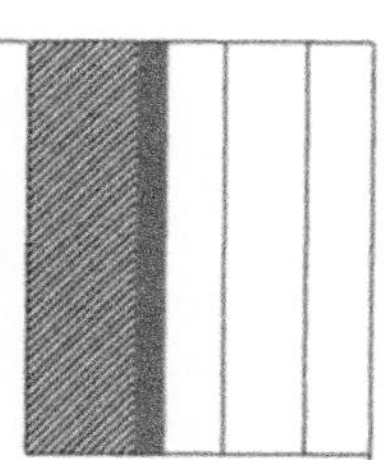

•

Section III: Hazard Prevention and Control
Conditional Goals (*Include cross reference to section, subsection, and question, e.g., I.B2*)
1.

90-Day Items (*Delete this section for final transmittal to National Office*)
1.

Best Practices
1.

Comments including Recommendations (*Optional*)
1.

Documents Referenced, Programs Reviewed (*Optional*)
1.

Section IV: Safety and Health Training	Yes or No	How Assessed: Interview	Observation	Doc Review

A. Safety and Health Training
A4. *Does the site provide minimally effective training to educate employees regarding the known hazards of the site and their controls? If not, please explain.* 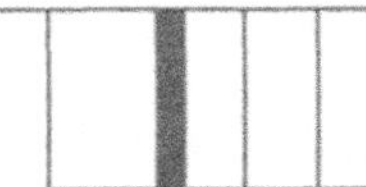
•

Section IV: Safety and Health Training
Conditional Goals (*Include cross reference to section, subsection, and question, e.g.. I.B2*)
1.

90-Day Items (*Delete this section for final transmittal to National Office*)
1.

Best Practices
1.

Comments including Recommendations (*Optional*)
1.

Documents Referenced, Programs Reviewed (*Optional*)
1.

11

VPP RESOURCES

As with any journey, it is always helpful to take advantage of the experiences of those who went before you. It is a much easier trek if you work with someone knowledgeable in the VPP process than try to find your own way. For those work sites interested in joining the ranks of VPP companies, there are a number of resources available to you that will provide valuable assistance.

THE VOLUNTARY PROTECTION PROGRAMS PARTICIPANTS ASSOCIATION (VPPPA)

The mission of the Voluntary Protection Program Participants Association (VPPPA) is "to be a leader in health and safety excellence through cooperation among communities, workers, industries and governments."

The VPPPA is the leading organization dedicated to cooperative occupational safety, health, and environmental management systems. The VPPPA, a nonprofit 501(c)(3) charitable organization, is a member-based association, providing a network of over 1900 companies and work sites that are involved in or in the process of applying to the Occupational Safety and Health Administration's (OSHA) or the Department of Energy's (DOE) Voluntary Protection Programs (VPP) and other government agencies that are developing or implementing cooperative recognition programs.

For over two decades, the VPPPA has provided its members a direct link to OSHA, DOE, and similar agencies within state plan states and continues to offer a support

Preparing for OSHA's Voluntary Protection Programs. By Brian T. Bennett and Norman R. Deitch
Copyright © 2010 John Wiley & Sons, Inc.

network for participants from a wide variety of industries. The VPPPA works closely with OSHA, OSHA state plan states, DOE, and the Environmental Protection Agency (EPA) in the development and implementation of cooperative programs within the agencies. Upon request, the association provides expertise in the form of comments and stakeholder feedback on rule makings and policies, as well as comments and testimony regarding legislation impacting worker health and safety and the environment.

The VPPPA provides occupational safety, health, and environmental leaders with networking and educational offerings, up-to-the-minute legislative information, industry advancements, preferred vendors and consultants dedicated to VPP, mentoring opportunities, professional development, and volunteer opportunities.

VPPPA MEMBERSHIP

As a nonprofit, member-driven association, VPPPA provides networking opportunities and educational offerings, publications, up-to-the-minute legislative information, mentoring, professional development, and volunteer opportunities.

Member sites include work sites approved for VPP by OSHA, state plan OSHA and DOE, sites pursuing VPP or looking to improve their safety, health and environmental programs, corporate offices, SHARP sites, and agencies and companies that provide promotional products and educational tools and services.

VPPPA CHAPTERS

VPPPA chapters are an established association presence in all of the 10 OSHA regions. Each chapter is headed by an elected board of directors that consist of safety and health leaders and both hourly and salary employees from union and non-union sites. Each chapter is governed by a set of bylaws that are modeled after the national VPPPA bylaws but are tailored to the needs and environments of the region.

The VPPPA chapter boards, committees, and volunteers are committed to providing local networking opportunities, educational sessions, VPP outreach, establishing mentoring relationships, and hosting annual conferences within each of the regions. Each VPPPA chapter works closely with its respective OSHA VPP Regional Managers, state plan contacts, and DOE VPP representatives.

Many of the VPPPA chapters have established regional websites, with detailed information about their programs, services, and deadlines. Links to the chapter websites can be found at www.vpppa.org along with other important information about each individual regional chapters.

THE VPPPA NATIONAL CONFERENCE

The VPPPA national conference is the premier forum for more than 3100 occupational safety and health environmental professionals, hourly workers, and managers. VPPPA conference participants represent a wide variety of individuals from VPP sites, sites seeking VPP approval, and organizations striving to improve their safety and health management systems. This four-day conference includes featured speakers from corporate America; two full days of workshops addressing current safety, health, and environmental issues; the VPPPA exhibit hall with over 200 vendors; and evening receptions and other social activities that provide the perfect opportunity for networking.

VPPPA conference participants represent professions including industrial hygienists, union representatives, consultants, environmental health specialists, and human resource managers. Government agency representatives from OSHA and DOE are available for networking in formal and informal settings.

The national conferences are usually held annually during the last week of August at different venues across the United States.

VPPPA MENTORING PROGRAM

The award-winning VPPPA Mentoring Program is a formal process to assist companies/facilities interested in the VPP or improving their safety and health management systems. The program matches interested sites with current VPP Star sites to help them achieve VPP recognition.

The association utilizes a network of regional coordinators and agency officials in establishing a mentoring relationship. Depending on preferences indicated on the VPPPA mentoring application form, coordinators consider similar experiences, industries, geographic location, and union representation when choosing prospective mentors.

Once a match has been made, the sites form a unique relationship based on the needs of the mentored site and the time and resources available to the mentor.

The VPPPA Mentoring Program is a free resource provided by the association to sites regardless of membership status.[1]

VPPPA ANNUAL AWARDS

The VPPPA annual awards for outreach and innovation recognize those VPPPA member sites and individuals that have made exceptional contributions to the mission of the VPPPA during the previous year. These pioneers in safety and health have shared their knowledge of safe practices with others or developed new techniques to create a safer, healthier work environment.

AWARD CATEGORIES

VPP Outreach Award

The purpose of the VPP Outreach Award is to provide recognition for "VPP Ambassadors." This award recognizes those who achieve an outstanding level of outreach activity and encourage others to share their knowledge. Reaching out to communicate and persuade others of the benefits of pursuing OSHA or DOE VPP approval is an important part of the mission of the VPPPA. Another part of the mission is to persuade other government regulatory agencies to adopt similar programs. The VPP Outreach Award is for an individual, company, or work site that has done an extraordinary amount of work in these areas.

Safety and Health Outreach Award

The purpose of the Safety and Health Outreach Award is to provide recognition for "VPP models," who reach out to share the safety, health, technical, and management expertise developed at their sites. The award is for an individual, company, or work site that has achieved an outstanding level of outreach in the safety and health arena, not directly encompassing the VPP.

VPP Innovation Award

The purpose of the VPP Innovation Award is to provide recognition for an individual, company, or work site that has developed and successfully implemented an innovation, encouraged others to try new approaches, and emphasized the value of creativity and flexibility in the resolution of worker safety and health problems. The innovation may be program related or of a technical nature.

These annual awards are presented at the annual national conference.

CONTACT THE VPPPA

For additional information about membership, the regional chapters, the annual conference, or mentoring, the VPPPA can be reached at:

VPPPA, Inc.
7600-E Leesburg Pike
Suite 100
Falls Church, VA 22043-2004
(703) 761-1146
www.vpppa.org

THE OCCUPATIONAL SAFETY AND HEALTH ADMINISTRATION (OSHA)

Cooperative Programs

OSHA offers the following cooperative programs under which businesses, labor groups, and other organizations can work cooperatively with the agency to help prevent fatalities, injuries, and illnesses in the workplace.

TABLE 11.1 Approved OSHA State Plan States

Alaska	New Mexico
Arizona	New York[a]
California	North Carolina
Connecticut[a]	Oregon
Hawaii	Puerto Rico
Indiana	South Carolina
Iowa	Tennessee
Kentucky	Utah
Maryland	Vermont
Michigan	Virgin Islands[a]
Minnesota	Virginia
Nevada	Washington
New Jersey[a]	Wyoming

[a]These state plans cover public sector (state and local government employment) only.

If you are located in a state with an OSHA-approved state plan (see Table 11.1), please contact your state agency for information about cooperative programs. All states with OSHA-approved programs have their own cooperative programs.

Alliance Program Through the Alliance Program, OSHA works with groups committed to safety and health, including businesses, trade or professional organizations, unions, and educational institutions, to leverage resources and expertise to develop compliance assistance tools and resources and share information with employers and employees to help prevent injuries, illnesses, and fatalities in the workplace. OSHA and the organization sign a formal agreement with goals that address training and education, outreach and communication, and promoting the national dialog on workplace safety and health.

Onsite Consultation Program Small and medium-sized businesses, particularly those in high-hazard industries or involved in hazardous operations, can use this free and confidential service to help improve their safety and health performance. Consultants from state agencies or universities work with employers to identify workplace hazards, provide advice on compliance with OSHA standards, and assist in establishing safety and health management systems. Onsite consultation services are totally separate from enforcement and do not result in penalties or citations.

Safety and Health Achievement Recognition Program (SHARP) Employers that have an onsite consultation visit with a complete hazard identification survey and meet other requirements may be recognized under SHARP for their exemplary safety and health management systems. Work sites that receive SHARP recognition are exempt from programmed inspections during the period that the SHARP certification is valid.

OSHA Challenge Pilot Program OSHA Challenge uses the VPP model as a foundation to provide a roadmap to eligible employers interested in implementing effective safety and health management systems. Volunteer administrators help guide the participants through a three-stage process. Graduates of OSHA Challenge receive recognition from OSHA and an invitation to apply for VPP with possible expedited approval. OSHA Challenge is available to general industry and construction employers in private and public sectors under OSHA's federal jurisdiction.

OSHA Strategic Partnership Program (OSPP) OSPP provides the opportunity for OSHA to partner with employers, employees, professional or trade associations, labor organizations, state onsite consultation projects, and/or other interested stakeholders. OSHA Strategic Partnerships are designed to encourage, assist, and recognize efforts to eliminate serious hazards and achieve model workplace safety and health practices. Each OSHA Strategic Partnership forms a unique, formal agreement that establishes specific goals, strategies, and performance measures. The OSPP is available to all private-sector industries and government agencies in locales where OSHA has jurisdiction.

Voluntary Protection Programs (VPP) VPP are OSHA's premier recognition programs for employers and employees who have implemented exemplary workplace safety and health management systems. A hallmark of VPP is the principle that management, labor, and OSHA work together in pursuit of a safe and healthy workplace. To attain VPP status, employers must demonstrate management commitment to the safety and health of their employees, and employees must be actively involved in activities that support the safety and health management system.

CONTACT OSHA

More information about OSHA's cooperative programs can be found at: http://www.osha.gov/dcsp/compliance_assistance/index_programs.html.

VPP DOCUMENTS

There are a number of documents that have been issued by OSHA that provide further details concerning the VPP.

Exhibit 11.1 is a copy of the Federal Register Notice (FRN 71800) to define new criteria intended to make the VPP more challenging and to raise the level of safety and health achievement expected of participants, and to bring the VPP's basic program elements into conformity with OSHA's Safety and Health Program Management Guidelines.

Exhibit 11.2 is a copy of the Federal Register Notice (FRN 120803) that revised the VPP in 2003. Exhibit 11.3 is a copy of the Federal Register Notice dated January 9, 2009, initiating revisions to the VPP including formalizing the Mobile

Workforce for Construction Demonstration Program and the Corporate VPP Pilot Program, as well as defining expectations for continuous improvement and outreach.

Exhibit 11.4 is a copy of CSP 03-01-003, commonly known as the VPP Policies and Procedures Manual. This document describes how OSHA administers the VPP with details about the program and responsibilities of all parties.

OSHA SPECIAL GOVERNMENT EMPLOYEE (SGE) PROGRAM

Once you have achieved VPP recognition, consideration should be given to allowing your employees to becoming part of OSHA's Special Government Employee Program.

The SGE program was established to allow industry employees to work alongside OSHA during VPP onsite evaluations. Not only does this innovative program benefit OSHA by supplementing its onsite evaluation teams, but it gives industry and government an opportunity to work together and share views and ideas.

Qualified volunteers from VPP sites are eligible to participate in the SGE program. These volunteers must be approved by OSHA and funded by their companies to participate. After submitting an application and completing the required training, these volunteers are sworn in as SGEs and are approved to assist as VPP onsite evaluation team members.

Figure 11.1 Brian Bennett receiving the OSHA National SGE of the Year Award from then Assistant Secretary of Labor Edwin Foulke, Jr. (Photo courtesy EHS Excellence Consulting Inc.)

As VPP grows, the support of SGEs will continue to be a critical component of the programs. The SGE program encompasses the spirit of VPP—industry, labor, and government cooperation. This cooperation embodies the idea of continuous improvement, which allows SGEs to bring a unique perspective to the team effort and take back to their sites ideas and best practices to further improve worker protections.

Many of the VPPPA regional chapters, as well as the national OSHA office, have developed an SGE of the Year Award to recognize those SGEs that go above and beyond to assist OSHA in its efforts to implement the VPP.

The VPP National Special Government Employee (SGE) Award recognizes an SGE who epitomizes and exhibits exceptional support, time, effort, and action in VPP, setting him or her apart from other SGEs. The person selected is actively involved in volunteer activities that benefit the VPP and its stakeholders and demonstrates outstanding commitment to the partnership ideals of the VPP. The first annual SGE of the Year was awarded by Assistant Secretary Jonathan Snare at the 2006 National Conference of the Voluntary Protection Programs Participants' Association (VPPPA). The award will be given on an annual basis.[2]

The author was honored to receive the National SGE of the Year Award in 2007 from Assistant Secretary of Labor Edwin Foulke (see Figure 11.1).

CONSULTANTS

Some facilities have decided the most cost-effective and efficient route to achieving VPP recognition is to secure the services of a consultant.

The key to success in selecting a consultant is finding one that has extensive experience with the program and specializes in assisting work sites achieve VPP recognition. By carefully selecting a consultant who is knowledgeable and experienced in the VPP, your resources can be focused on the specific tasks that must be completed in order to achieve VPP recognition. Although there are several consultants that provide VPP services, the premier company is EHS Excellence Consulting Inc. (www.ehs-excellence.com). EHS Excellence's staff includes a mix of former OSHA managers familiar with the administration and implementation of the VPP along with former private-sector employees from VPP work sites who have developed, implemented, and managed VPP Star caliber systems. These former private-sector employees include former Special Government Employees and hold numerous awards for their involvement with the VPP.[3]

REFERENCES

1. www.vpppa.org.
2. www.osha.gov.
3. www.ehs-excellence.com.

INDEX

Preparing for OSHA's Voluntary Protection Programs. By Brian T. Bennett and Norman R. Deitch
Copyright © 2010 John Wiley & Sons, Inc.

Printed and bound by CPI Group (UK) Ltd, Croydon, CR0 4YY

23/04/2025

14660906-0005